# The Philosophy of Ecology

Ecology is indispensable to understanding the biological world and addressing the environmental problems humanity faces. Its philosophy has never been more important. In this book, James Justus introduces readers to the philosophically rich issues ecology poses. Besides its crucial role in biological science generally, climate change, biodiversity loss, and other looming environmental challenges make ecology's role in understanding such threats and finding solutions all the more critical. Beyond its centrality to philosophers of biology, when ecology is applied and its insights marshalled to address these problems and guide policy formation, interesting philosophical issues also emerge. Justus sets them out in detail, and explores the often ethically charged dimensions of applied ecological science, using accessible language and a wealth of scientifically-informed examples.

JAMES JUSTUS is Professor of Philosophy at Florida State University. He has written numerous book chapters and journal articles on the philosophy of biology, the history of analytic philosophy, and general philosophy of science.

**Cambridge Introductions to Philosophy and Biology**

**General editor**

Michael Ruse, Florida State University

**Other titles in the series**

Derek Turner, *Paleontology: A Philosophical Introduction*

R. Paul Thompson, *Agro-technology: A Philosophical Introduction*

Michael Ruse, *The Philosophy of Human Evolution*

Paul Griffiths and Karola Stotz, *Genetics and Philosophy: An Introduction*

Richard A. Richards, *Biological Classification: A Philosophical Introduction*

Lynn Hankinson Nelson, *Biology and Feminism: A Philosophical Introduction*

Michael L. Peterson and Dennis R. Venema, *Biology, Religion, and Philosophy: An Introduction*

James Justus, *The Philosophy of Ecology: An Introduction*

# The Philosophy of Ecology

## An Introduction

JAMES JUSTUS

*Florida State University*

**CAMBRIDGE**
**UNIVERSITY PRESS**

University Printing House, Cambridge CB2 8BS, United Kingdom

One Liberty Plaza, 20th Floor, New York, NY 10006, USA

477 Williamstown Road, Port Melbourne, VIC 3207, Australia

314–321, 3rd Floor, Plot 3, Splendor Forum, Jasola District Centre,
New Delhi – 110025, India

79 Anson Road, #06–04/06, Singapore 079906

Cambridge University Press is part of the University of Cambridge.

It furthers the University's mission by disseminating knowledge in the pursuit of education, learning, and research at the highest international levels of excellence.

www.cambridge.org
Information on this title: www.cambridge.org/9781107040045
DOI: 10.1017/9781139626941



First published 2021

*A catalogue record for this publication is available from the British Library.*

*Library of Congress Cataloging-in-Publication Data*
Names: Justus, James, author.
Title: The philosophy of ecology : an introduction / James Justus, Florida State University.
Description: Cambridge, UK ; New York, NY : Cambridge University Press, 2021. | Series: Cambridge introductions to philosophy and biology | Includes bibliographical references and index.
Identifiers: LCCN 2020051646 (print) | LCCN 2020051647 (ebook) | ISBN 9781107040045 (hardback) | ISBN 9781107698154 (paperback) | ISBN 9781139626941 (epub)
Subjects: LCSH: Ecology–Philosophy.
Classification: LCC QH540.5 .J87 2021 (print) | LCC QH540.5 (ebook) | DDC 577.01–dc23
LC record available at https://lccn.loc.gov/2020051646
LC ebook record available at https://lccn.loc.gov/2020051647

ISBN 978-1-107-04004-5 Hardback
ISBN 978-1-107-69815-4 Paperback

# Contents

# Figures

# Preface

In his inimitable way, G. E. Hutchinson (1965) coined the phrase “the ecological theater and the evolutionary play.” At the time, the metaphor complemented other attempts to fortify the value of the synthesis of Mendelian genetics and Darwin’s theory (achieved just a few decades earlier) against the perceived hegemonic zeal of the new molecular and cellular approaches in biology (see Wilson 1994; Dietrich 1998). But the insight behind Hutchinson’s metaphor extends far beyond just the Modern Synthesis. It captures the indispensable contribution both ecology and evolutionary science make to understanding the biological world. Both concern vast albeit overlapping portions of that world, in both space and time, and alone neither can supply a complete accounting of it without the other. Just as ecological and evolutionary sciences are at the core of biology, philosophy of evolutionary biology *and* ecology are at the core of philosophy of biology. This book is an introduction to philosophy of ecology.

Beyond its ambition to explain vast portions of the biological world, and the crucial light it sheds on evolutionary dynamics, ecology has likely never been more important as a science, and its philosophy more important to society. Climate change, habitat degradation, biodiversity loss, and numerous other looming environmental challenges make ecology and the interrelated environmental sciences that help us understand the threats and identify feasible solutions all the more critical. When ecology is applied and its insights marshaled to address these problems and to guide environmental policy, interesting philosophical issues emerge. This book considers a few such issues in its final chapters.

This project has had a very long gestation, and throughout its development I have been unbelievably fortunate to learn from so many, in conversation and from their writings. They include Soazig Le Bihan, Mark Colyvan,

Chris Eliot, Alkistis Elliott-Graves, Paul Griffiths, Chris Lean, Stefan Linquist, Alan Love, Roberta Millstein, Jay Odenbaugh, Viorel Paslaru, Carl Salk, Carlos Santana, Elliott Sober, Derek Turner, Samantha Wakil, and Michael Weisberg, among many, many others. In addition, a significant portion of this book was completed while I was a visiting fellow at the Pittsburgh Center for Philosophy of Science in 2018. Thanks to Colin Allen, Michael Dietrich, Karen Kovaka, Edouard Machery, Sandy Mitchell, and Armin Shulz for especially stimulating discussion and helpful feedback.

Although most of the material in this book is new, some parts draw on previously published work. Chapter 1 is based on an article in *Philosophical Topics* (Justus 2019) and reprinted with the permission of the University of Arkansas Press. Chapter 4 draws on material from two articles published in *Philosophy of Science* (Justus 2008b, 2014), and Chapter 3 builds on a section of Justus 2013. Thanks to the University of Chicago Press and Springer Publishing Company for permission to use this material.

Last, two individuals deserve special recognition. First, Hilary Gaskin at Cambridge University Press has been unbelievably patient, supportive, and understanding from this project's inception so long ago to its culmination. I am profoundly appreciative. The same homage, to an even higher degree if that's possible, extends to Michael Ruse. Without a doubt, he is the sine qua non that made writing this book possible. He has been both unwaveringly supportive and constructively coercive, in precisely the right balance. I owe him immensely, will forever be grateful, and am lucky to be his colleague.

Finally, my greatest debt beyond my parents Jim and Lisa is to my wife, Staci, and two children, Ike and Aviva. Without the former and her constant encouragement, I never would have pursued let alone finished this work. Without the latter it would have been completed much faster, but with much less joy.

# Introduction

## Why Philosophy of Ecology?

Dobzhansky's sweeping generalization, "nothing in biology makes sense except in the light of evolution" (1964, 449), provocatively captures the centrality of evolutionary theory in contemporary biological science (see also Dobzhansky 1973). But his indelible rally call is also revisionist history, and grievously partial. Although the term "ecology" was not coined until 1866 (Haeckel 1866), most of what would be deemed biological investigation that did not concern the interior of organisms since at least the time of the ancient Greeks, and long before a nascent awareness of evolutionary forces, falls squarely within the purview of ecology. The biological understanding that laid the groundwork out of which evolutionary theory emerged was largely ecological.

Ecology therefore casts the same indispensable light in biology, and particularly on evolution. Ecological insights were an integral part of early evolutionary thinking; they are at the core of Darwin's original theory; and they will remain crucial to theorizing about how evolutionary dynamics shape the biological world. Consider evolutionary theory's central concept, natural selection. Evolution by natural selection is traditionally thought to depend on three population-level factors: phenotypic variation, heritability, and differential fitness (see Lewontin 1970).[1] All three are biologically crucial components, and at least the latter two have garnered significant attention from philosophers of biology (on heritability, see Tabery 2014 and Downes and Matthews 2019; on fitness, see Rosenberg and Bouchard 2015).

[1] Interestingly, population structure poses problems for this concise characterization of evolution by natural selection (see Godfrey-Smith 2007). Besides its relevance to population genetics, population structure is obviously also an important research topic within population ecology.

But without a doubt fitness is the conceptual and explanatory core of evolution by natural selection, and by far the most philosophical ink has been expended on it. What has not been recognized as widely or thoroughly as it should be is that fitness is a fundamentally ecological concept. Without wading into the substantial controversy about how exactly it should be characterized, it is safe to say that fitness depends on the relations between the traits of an organism and the various aspects of the environment it lives in. That is vague, of course, and hence the philosophical controversy. But studies of how organisms make their living in their different environments are about as central to ecology as it gets.

Moreover, indefensibly ignoring an ecological perspective is arguably responsible for flawed conceptions of fitness that motivate defining it in terms of reproductive rates, thereby abetting the infamous Popperian charge that evolutionary theory is tautologically vacuous. Properly *defining* fitness requires considering the organism–environment relations at the core of ecology. Only by disregarding those relations to focus exclusively on measures of reproductive success does the triviality threat gain purchase.

Fitness, in turn, is at the heart of other important biological concepts and explanations of biological phenomena, for example, adaptation, speciation, multilevel selection, niche construction, and perhaps even biological individuality, to name but a few. If fitness, evolution by natural selection, and evolutionary theory in general are unquestionably in the wheelhouse of any competent philosopher of biology, ecology should be as well.

Apart from its contribution to evolutionary theory, ecology also endeavors to account for vast portions of the living world directly. It is, for example, canonically characterized as the study of interactions between organisms and the environment. Its explanatory scope therefore includes not only these interactions but also their causal ramifications: the distributions and abundances of species they produce throughout the globe. Any science with an agenda this ambitious, especially one that pursues it with such sophisticated mathematical models and complex statistical methods for empirically testing them, deserves significant attention from philosophers of science.

Thus far the title question has received two answers: (1) an ecological perspective underpins much of evolutionary theory, so competent philosophy of the latter, whose value is unquestioned, requires the same of the former; and (2) any science with such a global scope merits philosophical attention. Answer (2) is generic. Many other sciences are in the same camp, and like ecology, many are beginning to receive appropriate interest from philosophers of science: archeology (Chapman and Wylie 2016), chemistry (Hendry et al. 2011), paleontology (Turner 2011; Currie 2018), and others. Answer (1) is specific to ecology, and perhaps a few other biological disciplines that contribute directly to understanding evolutionary dynamics, for example, developmental and molecular biology. But answer (1) is also derivative. Philosophy of ecology's significance piggybacks on the philosophical significance of evolution. Dependency is not diminishment, but the value is not autonomous.

Fortunately, there is ample autonomous value to go around. For starters, the systems studied in ecology are unbelievably complex. Ecosystems contain a plethora of distinct kinds of entities, which interact in an untold number of ways, and do so on numerous spatial and temporal scales. Even the simple task of representing these systems in a model poses interesting philosophical issues about, for example, the nature and justification of (usually necessary) idealizations (Weisberg 2007), when inference under conditions of significant uncertainty is reliable (Justus 2012a), how the epistemic credentials of such (sometimes quite unwieldy) models can be evaluated (Winsberg 2018), the ultimate limits of representations in science (van Fraassen 2008; Weisberg 2013), and many others.[2]

The magnitude of this complexity does not mean it is necessarily unmanageable or that simple unifying principles will never be found. Imagine the natural philosopher-scientist well before Mendeleev and his predecessors. It understandably would have seemed preposterous that the seemingly innumerable types of substances exhibiting such a vast array of different properties were actually composed of only a relatively small number of elements, elements that in turn could be grouped into an even smaller number of categories that explained much of their nature. Despite its apparently dismal odds, the periodic table was created and this unruly diversity tamed.

But the stubborn fact is that an analogous ecological periodic table has not been uncovered, after a century and a half of continuous scientific inquiry since Mendeleev's breakthrough, and with cutting-edge technologies and statistical methods of data analysis that far exceed the investigative capabilities of anything nineteenth-century scientists could have imagined, let alone

[2] For an excellent overview of all these issues in an ecological context, see Odenbaugh 2019.

have had access to. We should not be surprised. The chemical world is complicated, but at the elemental level its compositional complexity is much less daunting. Chemical processes are also much more tightly tethered to numerous law-like regularities, conservation of matter, conservation of energy, principles of electromagnetism, and so on. These regularities govern ecological systems as well, of course, but the constraint they impose is much more slack. Biological communities as diverse as the Amazon rainforest, Saharan desert, and Siberian boreal forests all dutifully conform to the law-like regularities, but these and other epistemic triumphs from physics offer little to explain the stark ecosystem differences.[3]

With little mooring in physics below and no grand unified theory governing from above, economics, rather than physics or chemistry, seems to constitute a close disciplinary analog to ecology (see Shulz 2020). Economics, like ecology, trades in extremely sophisticated mathematical models. And, like ecology, it lacks anything remotely resembling a comprehensive *and plausible* theoretical framework.[4] And perhaps because both disciplines lack such a framework they often utilize concepts and methods developed in other fields, such as physics, to construct those sophisticated mathematical models, which itself raises interesting philosophical issues (see Justus 2008b). Although economic data are plentiful, acquiring the kind of data that would definitively confirm or disconfirm economic models is extremely difficult. The same challenge confronts ecological modeling. Being heavy mathematically and light on (relevant) data makes economics and ecology philosophically rich subjects in their own right (see Kincaid and Ross 2017).

There is another dimension to the philosophical significance of ecology. Sciences are human activities that occur in broader cultural and societal

[3] It should be stressed that the challenge complexity poses, which differentiates ecology from many but certainly not all other sciences, does not establish some *inherent* difference, that it possesses some fundamentally distinct, autonomous nature that necessitates different epistemological approaches and methodologies from most other sciences. The complexity of ecological phenomena partially explains the current epistemic and methodological character of the science, and that dependency is itself philosophically interesting. But it is far from establishing ecology as an autonomous special science, whatever that may mean. Chapter 4 in fact shows ecology can benefit greatly from practices developed in other sciences if deployed wisely, mathematical models being the case study. For a sustained critique of the autonomy line for biology and defense of the role mathematics has in unifying the sciences, see Thompson 1995.

[4] Here I follow behavioral economists in holding that the death knell of *Homo economicus* and the Chicago school rattled long ago (see Kahneman and Tversky 2000).

contexts.[5] The latter always bears on the former, and the former sometimes on the latter, but not nearly with the same intensity across disciplines. The newest discoveries in carbon sequestration or cancer sequencing have a social potency that the latest findings in vampire bat mating strategies or astrogeology do not. Ecology is unmistakably toward the potent end of that spectrum. As the impacts humans have on the natural world magnify, ecological studies can reveal their full, horrific ramifications. And ecology's expansive investigative purview ensures it is uniquely positioned to expose the details of those ramifications across a wide variety of different kinds of ecosystems and spatial scales, as well as possibly identify how they can be mitigated. In a way, ecological knowledge might provide the scientific antidote to ultimately catastrophic societal tendencies.

Similarly, ecology has uniquely epistemic authority vis-à-vis environmental ethics. First, the revelatory function described above can assist ethical theorizing. Ecology furnishes scientific facts that ethicists must recognize and respond to. The details of how global warming will affect coastal communities, for example, raise daunting issues about inequity and the environmentally exacerbated ramifications of economic and political inequality. Independent of concerns about animal welfare, the ecological effects of factory farming should also inform ethical judgments about them.

But there is a second kind of link between the disciplines. Environmental ethics often trades in concepts and claims that have both normative and descriptive content. For example, whether a negative ethical appraisal of exotic or invasive species is defensible depends on what they are and what they are capable of (Elliot-Graves 2016). Ethicists alone cannot answer those questions; ecological input is essential. Sober's (1986) trenchant criticism of environmental ethicists' use of a "naturalness" concept showcases the salience of biological science, ecology in particular, in environmental ethics.[6] That input might influence the valence of an ethical judgment, or clarify that one basis for an ethical position is inferior to another or outright indefensible. Whether there is a

[5] Here it is crucial to sharply distinguish ecology the science from popular characterizations of the term 'ecology' associated with environmental and other sociopolitical views, such as that "everything is interconnected" or new-age versions of the Gaia hypothesis (see Ruse 2013). For those with little exposure to biological science, the two connotations are often conflated.

[6] "[T]o the degree that 'natural' means anything biologically, it means very little ethically. And, conversely, to the degree that 'natural' is understood as a normative concept, it has very little to do with biology" (p. 180).

"balance of nature" as Paul Taylor (1986) intends that phrase is another example of where ecological science should bear on theories of value in the natural world.

So far, I have described general features of ecological science that garner philosophical interest. But, as with any new and burgeoning field, focus has congealed around several broad areas:

1. conceptual issues in the history of ecology
2. characterizing problematically unclear ecological concepts, especially "biodiversity" and "stability"
3. whether there are distinctively ecological laws
4. reduction in ecological science and the reality of biological communities
5. the role of mathematical modeling in ecology
6. the relationships between evolutionary theory and ecology, and conservation science and ecology.
7. the role of non-epistemic values in applied sciences

Beyond a narrow focus on ecology, some of these areas offer novel insights into standard topics in general philosophy of science, such as emergence and reduction, the nature of laws of nature, conceptual content and concept determination, the status and function of models in science, and the status and function of values in sciences.

Others areas involve topics unique to ecology, and to which philosophers can make valuable contributions to scientific practice. Each area, in turn, covers numerous specific issues. With respect to item (4), for example, some ecologists and philosophers of science have recently proposed an analogy between Newtonian mechanics and ecosystem dynamics (Ginzburg and Colyvan 2004). Although the status and epistemic utility of this analogy remain controversial, this work suggests a close parallel should exist between modeling strategies in physics and ecology. But other analyses counter this parallel. For example, Hubbell's (2001) unified neutral theory of biodiversity primarily derives from theories developed within biology proper: MacArthur and Wilson's (1967) theory of island biogeography and Kimura's (1983) neutral theory of molecular evolution. And one concept of stability appropriated from physics and often employed in ecological modeling, Lyapunov stability, seems unable to capture the ecological phenomena it is intended to represent (Justus 2008b). Analyses of this unresolved issue shed light on the different role that models may have in biology and physics in general. With respect to item (1), to cite another prominent example, there are several concepts besides "biodiversity" and "stability" central to ecological science and in need of conceptual clarification, including "carrying capacity," "community," "complexity," "disturbance," "ecosystem," "habitat," "keystone species," "niche," "population," and many others. Like most concepts in developing sciences, fully adequate definitions of these and other ecological concepts have not yet been formulated. These and other issues provide rich conceptual grist for philosophers of ecology.

As these examples illustrate, ecology concerns a diverse conceptual terrain and an interesting set of theoretical and methodological issues, thus far underexplored by philosophers of science. The subsequent chapters describe its main contours and introduce readers to some of the most exciting topics in philosophy of ecology.

Chapter 1 scrutinizes the ecological unit thought to underlie the structure of biological communities and perhaps provide a "conceptual foundation" for the science: the niche. The history of the concept's origin and development is recounted, from its beginning with Joseph Grinnell and Charles Elton, and culminating in G. E. Hutchinson's highly abstract *n*-dimensional hyper-volume account. The niche is widely believed to be a fundamental abstraction in ecological theorizing, essential to ensuring its generality. For example, general accounts of the similar structure of communities composed of different species are only possible, it seems, if a shared underlying niche structure generates the similarity. Grasslands in the central plains of North America and Africa share a similar structure and exhibit similar dynamics because they instantiate roughly the same system of niches, it is claimed, albeit with different species. This is only one of many seemingly indispensable functions of the niche concept. Appeals to niche structure seem to provide the only explanation of convergent evolution, character displacement, as well as evolutionary convergence of ecosystems: remarkably similar biological communities emerging over geologic time scales (e.g., past communities with saber-tooth tigers as apex predators and present communities with *Canis* species functioning similarly).

But this paradigm has been challenged in at least two ways. First, "neutral" theories of community structure, particularly Stephen Hubbell's "Unified Neutral Theory of Biodiversity," pose a serious threat to the putative indispensability of niche thinking. By emphasizing the role of dispersal limitation, sampling effects, and stochasticity within a cohesive model of community dynamics, neutralists have formulated a cogent alternative to "rules of community assembly" based on niche structure. Second, niche constructionists' recent charge that many niches are made, not found, seems to make the standard account inapplicable. If organisms can modify their environments and thereby their niches to increase fitness, it is no longer clear the niche has explanatory priority. What explains community structure, convergent evolution, character displacement, and the like is no longer an extant niche structure that specific communities realize or that imposes a selection regime producing convergence and displacement. Rather, a locus of explanatory force resides within organisms that do the niche constructing. After carefully examining the content of proposed definitions of the niche, and the supposed contributions it makes to ecology theory, Chapter 1 also arrives at a negative assessment, but with a very different basis. Despite its supposed centrality, the analysis surprisingly concludes the niche concept is dispensable. It simply does not do the significant conceptual or explanatory work in ecology it is claimed to do.

Chapter 2 connects ecology with two central issues in general philosophy of science: what marks the real, and the nature of laws of nature. On the former, biological communities are the problem case. The question is whether they are anything more than the individual organisms of different species comprising them. If they are not, presumably they possess no independent existence. If they are, an account is needed of (1) this "something more" *and* (2) how it confers independent existence. Absent either, realist aspirations are frustrated. The first task requires a careful dissection of community structure, community dynamics, species distribution patterns, and what they reveal about how groups of species might assemble into communities. For example, individualists claim that species distributions along environmental gradients overlap continuously and significantly, and do not form discrete boundaries. But, the argument continues, communities are only real if they have such distinct boundaries. So they are not real. The second task involves delving into metaphysics, principally to determine whether the "something more" these ecological assemblages possess actually "cuts nature at its joints," the proverbial criterion for ontological credibility according to realists. These issues have catalyzed lively debate in several recent publications, and vetting the arguments contained therein is one of the two main goals of this chapter.

The second goal is addressing a similarly intricate and fundamental topic: whether there are laws in ecology. There are various challenges to the idea that there are such laws: ecology's relative paucity of predictive success, that its models and experimental results lack sufficient generality, that candidate laws are riddled with exceptions, and that ecological systems are too complex. But complexity is surely a surmountable obstacle. It is difficult to imagine a more complicated system than the entire cosmos, but no one suggests its complexity is not governed by relativistic and quantum mechanical laws, or that humans do not continue to uncover their form. Some philosophers have recently argued that other properties thought to preclude a discipline from trading in laws – limited predictive accuracy, generality, not being exceptionless – should be jettisoned, and that ecology indicates why. Ecology, they argue, has uncovered regularities, such as Gause's supposed law of competitive exclusion and numerous allometries, that possess a kind of necessity and therefore merit the label law of nature.

The idea there is a "balance of nature" in focus in Chapter 3 was a staple of the schools of natural philosophy from which biology emerged, long before the term "ecology" was even coined. Some early ecologists continued this tradition by attempting to derive the existence of a "natural balance" in biological populations from organismic metaphors and anthologies with physical systems. Not until the second half of the twentieth century was the concept of a balance of nature rigorously characterized as a kind of stability, and the predominantly metaphysical speculations about its cause superseded with scientific hypotheses about its basis. But significant uncertainty and controversy exists about what features of an ecological system's dynamics should be considered its stability, and thus no consensus has emerged about how ecological stability should be defined. Instead, ecologists have employed a confusing multitude of different terms to attempt to capture apparent stability properties: "constancy," "persistence," "resilience," "resistance," "robustness," "tolerance," and many more. This, in turn, has resulted in conflicting conclusions about debates concerning the concept based on studies using distinct senses of ecological stability.

Different analyses seem to support conflicting claims and indicate an underlying lack of conceptual clarity about ecological stability that this chapter diagnoses and resolves. In particular, a comprehensive account of stability is presented that clarifies the concepts ecologists have used that are defensible, their interrelationships, and their potential relationships with other biological properties, including diversity and so-called ecosystem functioning. Chapter 3 also evaluates the intriguing idea developed by some ecologists and philosophers of science that there *must* be a balance of nature given the claimed necessity of density-dependence and the negative feedback mechanism it imposes on population growth. Besides providing insights about how problematic scientific concepts should be characterized, it is worth noting that the issues addressed in this chapter have a potential bearing on biodiversity conservation. It seems that for most senses of stability, more stable communities are better able to withstand environmental disturbances, thereby decreasing the risk of species extinction. If there is a systematic positive feedback between diversity and stability, that would therefore support conservation efforts to preserve biodiversity.

To learn anything significant about the natural systems, ecologists have to represent them. The most common types of representation in ecology, and science in general, are mathematical models. Models of biological populations and communities take a wide array of functional forms and can contain many different types of variables and parameters. This complexity, the focus of Chapter 4, makes for fertile philosophical fodder and connects ecology to the extensive literature in general philosophy of science on modeling and scientific representation. To manage this complexity, ecologists sometimes borrow concepts and methods developed in other sciences. The fruits and perils of such cross-disciplinary fertilization is explored with two case studies: the methodological individualism of individual-based models, which connects ecology and social science, and defining ecological stability as Lyapunov stability, which connects ecology and physics. The first connection bears significant fruits, the second proves perilous.

Some biological communities are clearly more complicated than others. For example, tropical communities usually contain more species; there is evidence their species interact more intensely; these interactions are more variegated in form; and they exhibit more trophic levels than high latitude communities. Ecologists often invoke the concept of diversity to represent these differences in the "complicatedness" of communities: tropical communities are often said to be more ecologically diverse than tundra communities. Chapter 5 explains how "biodiversity" (coined as a simple shorthand for "biological diversity" in the mid-1980s) captures this notion of ecological diversity and much more, including developmental, morphological, and taxonomic diversity. Simply put, it designates the diversity of biological systems at all organizational levels, the population and community levels being the most common focus in ecology. How biodiversity should be characterized therefore depends on how these systems are represented, particularly on how their parts are individuated, classified, and distributed among those classes. Representations may vary with different explanatory or predictive scientific goals, and across types of systems, so characterizations of biodiversity may vary across these contexts as well.

Philosophers are drawn to the concept of biodiversity given its problematic complexity and the interesting theoretical and methodological issues the sciences studying and endeavoring to protect it raise. Its significance is common currency within environmental ethics, but biodiversity has only recently garnered broader attention from philosophers of science. This chapter describes the main contours of the concept and guides the reader through some of the growing scientifically oriented philosophical literature on biodiversity. It also makes the connection, explored in great detail in Chapter 6, between ecology proper and the kind of applied ecology conducted in efforts to conserve biodiversity, conservation biology.

Chapter 6 shows that the notion of progress for ethically driven applied sciences needs to be rethought. Conservation biology emerged as a rigorous science focused on protecting biodiversity, and as a discipline of applied ecology distinct from pure ecology, in the 1980s. Two algorithmic breakthroughs in information processing made this possible: place-prioritization algorithms and geographical information systems. They provided a defensible, data-driven methodology for designing reserves to conserve biodiversity. This obviated the need for largely intuitive and highly problematic appeals to ecological theory to design reserves at the time. They also supplied quantitative, largely critical assessments of existing reserves. Most reserves had been designated on unsystematic, ad hoc grounds and consequently poorly represented biodiversity. Demonstrating this convincingly was unsurprisingly crucial to ensuring biodiversity would be adequately protected in future policy-making contexts.

Despite these unquestionable advances, the notion that they constitute scientific "progress" has recently been criticized. Traditional ecological theory, such as island biogeography theory, it is claimed, is required for genuine progress about reserve design; algorithmic innovation in data processing is insufficient. Place-prioritization algorithms are also supposedly less scientifically grounded and produce reserves that poorly protect biodiversity. Chapter 6 argues that on all accounts this criticism is indefensible and involves numerous inaccuracies about the science, misconstrues the character of applied science, and relies on an untenable conception of progress for applied sciences with ethical objectives such as conservation biology. Although applied sciences are unquestionably science and employ scientific methods, what constitutes progress within them should not always be judged by the standards of classic descriptive sciences such as chemistry, evolutionary biology, and physics.

Chapter 7 attempts to clarify the *proper* role of ethical values in ethically driven disciplines of applied ecology such as conservation biology, invasion biology, and restoration ecology. Most sciences are principally concerned with discovering and explaining phenomena, but applied sciences sometimes have a different, explicitly ethical agenda. Some applied sciences pursue more immediately pressing goals, such as solving societal problems. Applied ecology and biodiversity conservation, and medical science and human health are two examples. Nonepistemic values concerning ethical goals seemingly permeate these "teleological" sciences. One of the most direct ways in which ethical and sociopolitical values bear on ecology (and vice versa) is in population viability analyses (PVAs). These are studies, usually model-based, of the dynamics of biological populations and how they would respond to various disturbance and management regimes. Whether the data are sufficient to show a regime adequately ensures a stipulated viability threshold usually requires a trade-off between minimizing type I and type II errors. This in turn seems to require the input of nonepistemic, ethical values. As such, PVAs have a significant bearing on conservation planning and action and seem to essentially incorporate ethical assumptions and considerations. Choices of scientific categories and terms, such as "carcinogen" and "endangered," seem to be similarly infused with ethics. Numerous other examples could be cited.

This influence has recently encouraged the view that they are value-laden in a strong sense: both ethical values and nonnormative facts contribute indispensably to teleological science, and their respective contributions cannot be demarcated. In fact, the inextricable suffusion of value supposedly challenges a clear fact/value distinction. Some have also argued that this influence begets an unacceptable relativism in scientific testing in applied ecology: which hypotheses are ultimately accepted or rejected will be determined by the ethical evaluation of the relevant states of affair, such as whether species conservation is worth doing. Chapter 7 describes these charges but also argues they are overstated. The value-laden character of these sciences does not challenge the fact/value distinction or the objectivity of hypothesis testing in applied sciences. Rather, although ethical values influence the general structure and methodologies of applied ecology, these influences can be demarcated from the factual status of claims made within it.

# 1 The Ecological Niche

Perhaps no concept has been thought more important to ecological theorizing than the niche. Without it, technically sophisticated and well-regarded accounts of character displacement, ecological equivalence, limiting similarity, and others would seemingly never have been developed. The niche is also widely considered the centerpiece of the best candidate for a distinctively ecological law, the competitive exclusion principle (see Chapter 2 on ecological laws).

The received view in ecology has therefore been that the niche is indispensable, despite occasionally vocal protests from a small minority. After all, the concept is often said to simply explicate the idea that species make their biological livelihoods in different ways, and what could be more central to ecology? Many (if not most) influential analyses in the 1960–1970s bore the "niche" label, often paying homage to Hutchinson's (1957) highly abstract definition in particular. Mechanistic models of resource consumption that predominated subsequent decades (e.g., Tilman 1982) are taken to extend and refine the same approach, the niche similarly at their core. More recently, Hubbell's (2001) unified neutral theory certainly perturbed the prevailing assessment, but it fell far short of upending it (see Odenbaugh forthcoming). A prominent book responding to neutralist theories in favor of niche-based theorizing, for instance, proclaims that "the niche has provided and can continue to provide the central conceptual foundation for ecological studies" (Chase and Leibold 2003, 17).

In this case, however, the naysayers were right. The incongruous array and imprecise character of proposed definitions of the concept square poorly with its apparent scientific centrality. Rather than reflect innocuous semantic differences or a potentially useful integrative pluralism, this definitional diversity and imprecision reflects a problematic conceptual indeterminacy that challenges its putative indispensability in ecology. The niche has not and cannot – at least as it has been characterized thus far – do the substantive, foundational work it is claimed to do in ecology. The conceptual content tethered to the term "niche" is just too problematically disjoint and amorphous to play that role. Unsurprisingly, although the term frequently received lip service, the content specified in its proposed definitions actually contributed little to the analyses mentioned above.

This gap between conceptual aspiration and scientific practice permeates appraisals of ecological theorizing to the present in many different, multifaceted ways. This chapter focuses specifically on the concept's origins in the work leading up to and in many ways culminating in Hutchinson's highly abstract *n*-dimensional hypervolume definition. The same kind of shortcoming is exhibited throughout this development, for different reasons. Section 1 describes the emergence of the ecological niche in Joseph Grinnell and Charles Elton's work. From the very beginning, the concept's content was unmistakably disjoint: environments and how they impact species was one focus; how species function in communities, particularly via trophic interactions, was the other. Beyond the bivalent focus, the concept was also problematically imprecise. This point is illustrated by considering the contentious idea of "vacant" niches and the significant indeterminacy about their possibility.

Despite the widespread view that the niche concept is the centerpiece, Section 2 argues the niche concept is not the centerpiece of perhaps the best candidate for a distinctively ecological law, the competitive exclusion principle. (See Chapter 2 on ecological laws for an analysis of whether the principle should be considered a law.) Gause's (1934) paramecium experiments and putatively mechanistic explanation of competitive dynamics with Lotka–Volterra equations are widely taken to supply the first compelling grounds for a niche-based version of the exclusion principle (see Hutchinson 1978). But the evidence for this judgment is pretty thin. Gause's explanation actually makes little use of niche ideas. His semantically suggestive – but thoroughly non-niche – term "vacant places" may have led many later commentators astray. And what little he did say about niches provides little guidance about how niche considerations might be brought to bear on models of competitive dynamics, via competition coefficients, for instance. This serious deficiency is a quite general problem, one shared by later attempts to characterize the concept, including the most influential definition of the concept in ecology.

That definition is Hutchinson's *n*-dimensional hypervolume characterization of the niche, the focus of Section 3. As judged by the attention it received, his definition had an enormous impact on ecology. But judged by the content conveyed, the impact seems disproportionate. Hutchinson made highly questionable and significantly limiting assumptions in characterizing the concept, and at least one conceptual confusion seems to render the concept quite intractable. The most serious deficiency, however, is the same kind of conceptual indeterminacy exhibited by earlier attempts to pinpoint the concept. Chase and Leibold's (2003) more recent revival of the niche is problematic in the same way. Rather than convey information about community dynamics, information that helps represent and analyze those dynamics, the niche superfluously supervenes on them on their account of the concept.

Problems with the concept aside, the actual work done under the niche rubric could be useful, and frequently was. Being able to define the concept, or even reliably identify species' niches, is not very important. What is important, potentially, is recognizing possible patterns across biological systems: even evolutionarily distant communities in different geographical regions sometimes (perhaps often) realize a similar causal structure. That commonality can then serve as a basis for extrapolation and generalization. Section 4 concludes by describing a defensible kind of community-level – as opposed to species-level – causal inference that might underlie Elton's (and perhaps Grinnell's) niche discussions. It also raises a general issue about how the utility of concepts should be gauged in science.

## 1 Grinnell and Elton's Niches

The first director of the Museum of Vertebrate Zoology at the University of California at Berkeley, Joseph Grinnell, was not the first to use "niche" in an ecological sense,[1] but he was the first to do so with any significance. His most well-known paper doing so – the first ecological publication with "niche" in the title – contains three instances all in the penultimate paragraph:

> These various circumstances, which emphasize dependence upon cover, and adaptation in physical structure and temperament thereto, go to demonstrate the nature of the ultimate associational niche occupied by the California

[1] Apparently, Roswell Johnson was in a 1910 analysis of ladybug color patterns (see Hutchinson 1978, 155–156).

> Thrasher. This is one of the minor niches which with their occupants all together make up the chaparral association. It is, of course, axiomatic that no two species regularly established in a single fauna have precisely the same niche relationships. (Grinnell 1917, 433)[2]

The "circumstances" are the chaparral habitat's physical characteristics, for which the thrasher's phenotypic properties are especially well suited. Its dense undercanopy foliage prevents all but short bursts of flight, complementing the thrasher's small, compact wings. Its inconspicuous drab-brown plumage also enhances predator evasion in that foliage.

The allusions to occupation are important. Grinnell's primary focus was animal species; the vegetation they inhabit was conceptualized as part of their physical environments. Niches are then units of that physical, partly biotic environment for Grinnell, units species can occupy. With evolutionary history in mind, relationships between occupants and what is occupied can therefore be explanatory: properties of niches, as actual bits of physical space, can account for why organisms residing in them possess the (adaptive) phenotypic properties they do.

Although a minority view, there are contrasting readings of Grinnell. Hutchinson (1978, 157) claimed, "it is evident that [for Grinnell] the space occupied by 'just one species' is an abstract space that cannot be a subdivision of the ordinary habitat space." But such abstraction coheres poorly with Grinnell's extensive descriptions of actual portions of environments as niches. The emphasis on the concrete is quite clear in later papers:

> Habitats have been variously classified by students of geographical distribution. Some of us have concluded that we can usefully recognize, as measures of distributional behavior, the realm, the region, the life-zone, the fauna, the subfauna, the association, and the ecologic or environmental niche. The latter, ultimate unit, is occupied by just one species or subspecies; if a new ecologic niche arises, or if a niche is vacated, nature hastens to supply an occupant, from whatever material may be available. Nature abhors a vacuum in the animate world as well as in the inanimate world. (Grinnell 1924, 227)

[2] Note the last sentence's close similarity with what was later labeled the "competitive exclusion principle" (see Section 2).

Note that *habitats* are being classified with different measures of *geographical distribution*, the unit of finest resolution being the "ecologic or environmental" niche (synonymy implied).[3] The idea that an abstraction is really what is being invoked therefore appears implausible. That Hutchinson favored and developed an abstract-space approach himself may be relevant.

In one of ecology's founding works, *Animal Ecology*, Charles Elton (1927, 63–64) described a very different concept:

> Animals have all manner of external factors acting upon them – chemical, physical, and biotic – and the "niche" of an animal means its place in the biotic environment, *its relations to food and enemies*. The ecologist should cultivate the habit of looking at animals from this point of view as well as from the ordinary standpoints of appearance, names, affinities, and past history. When an ecologist says "there goes a badger" he should include in his thoughts some definite idea of the animal's place in the community to which it belongs, just as if he had said "there goes the vicar."

Elton recognized the impacts of abiotic ("chemical," "physical") factors, but unlike Grinnell his niche focuses on biotic interactions. The one diagram presented in the "Niches" section, for example, illustrates the "niche occupied by small sapsuckers, of which one of the biggest groups is the plant-lice or aphids" (1927, 66), with a "food cycle" – "food web" in current terminology – comprised solely of biological nodes (Figure 1).

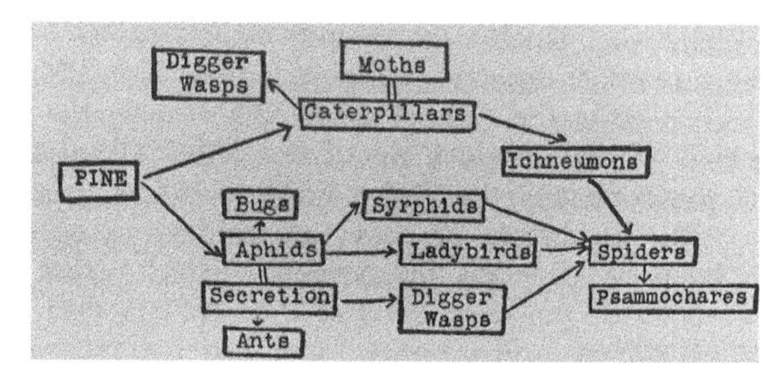


Figure 1 Elton's diagram of a "food cycle" intended to illustrate the niche of the "small sapsuckers." From Elton 1927, 66

[3] "Fauna," "subfauna," and "association" were not categories of biological composition for Grinnell. Rather, they were hierarchical units of geography – determined primarily by abiotic factors such as humidity and temperature – within which species distributions could be classified and hypotheses about their causes evaluated (Griesemer 1990).

This figure and reference to a "small sapsuckers" niche (note the plural) reveals an interesting aspect of Elton's concept absent from Grinnell's. For Elton, what individuates niches partially depends on how fine-grained the relations between organisms comprising a biological community are conceptualized. On this point Elton (1927, 64) was quite explicit:

> [W]e might take as a niche all the carnivores which prey upon small mammals, and distinguish them from those which prey upon insects. When we do this it is immediately seen that the niches about which we have been speaking are only smaller subdivisions of the old conceptions of carnivore, herbivore, insectivore, etc., and that we are only attempting to give more accurate and detailed definitions of the food habits of animals.

Degree of representational resolution therefore determines what counts as a niche. And since that degree differs across scientific contexts, species are constituents of many distinct niches with different extensions. The claim that coexisting species cannot have the same niche – one gloss on the competitive exclusion principle that Grinnell expressed in the first quote above – would thus sound foreign to Elton. It relies on a much narrower niche conception. This greater generality may account for the fact, noted by Hutchinson (1978, 152), that Elton never used the niche concept to explain competition.[4] It likely also explains why he was never tempted to elevate competitive exclusion to the axiomatic or principle status.

Dependence on representational resolution therefore marks an important contrast between Grinnell's and Elton's understandings of niche. But the contrast also seems to stem from a much deeper and more significant divergence of habitat versus functional conceptions. Unlike Grinnell's markedly physical conception, Elton's vicar analogy manifests the latter. Just as vicars are identified by functional roles in religious institutions, an animal's "place in the community" is similarly functionally individuated by its role in networks of biotic interactions.

[4] It should be highlighted that Elton *could* have attempted such an explanation. The more precisely the functional relations being considered under the "niche" label are described, the more likely those relations could reveal competitive dynamics. For example, precisely describing the food habits of hyenas and lions would reveal they are competitors. But it then seems the description of those habits – as captured in explicit competition equations – is what yields the insight. The question is what the niche concept contributes (see below and Section 2).

Some resist this bipartite judgment (e.g., Schoener 1989; Griesemer 1992). Griesemer rightly stresses that Grinnell and Elton had different research foci – primarily, how evolutionary dynamics influence species distributions versus how trophic interactions determine community structure – and that that probably partially explains their divergent uses of "niche." He also correctly highlights that each recognized the importance of biotic *and* abiotic factors, while obviously believing one was more salient than the other. The claim that Elton and Grinnell possessed distinct biotic and abiotic concepts should therefore be rejected. These differences, though nontrivial, only indicate different conceptual emphases for Griesemer, not different concepts: "Grinnell and Elton both identified the niche as the place/role a species happens to occupy in an environment" (1992, 235).

Correctly judging when differences in conceptual emphasis signify distinct concepts requires a theory of concept individuation, which is beyond the purview of Griesemer's analysis (or this one). But some of the dissimilarities seem to make concept-individuating differences. For instance, the clause "occupy in an environment" in Griesemer's characterization is problematic in Elton's case. Elton did mention an animal's place in the "biotic environment" – though just once in *Animal Ecology* (see above) – but the immediate, emphasized paraphrase ("*its relations to food and enemies*") strongly suggests the functional sense given explicitly two sentences later in the vicar analogy with the very similar phrase "animal's place in the community."

Elton's recognition of a small-mammal-consuming-carnivore niche or carnivore niche in toto reinforces this conclusion. These classes are characterized not in relation to environments their members occupy, but rather by their functional role in communities: consuming fleshy prey. On this issue Elton was unambiguous. The "Niches" section begins, "[A]lthough the actual species of animals are different in different habitats, the ground plan of every animal community is much the same. In every community we should find herbivorous and carnivorous and scavenging animals." After giving specific examples the paragraph continues, "It is therefore convenient to have some term to describe the status of an animal in its community, to indicate what it is *doing* and not merely what it looks like, and the term used is 'niche'" (1927, 63). Niches are thereby characterized in terms of this common "ground plan": an abstract pattern of basic functional relations underlying all communities according to Elton. But that plan is contrasted with the different habitats species inhabit. Irrespective of whether niches are "places" or "roles" – two terms that often have very different connotations – for Elton they are not parts of environments. This concept, unlike Grinnell's, is thoroughly functional.[5]

There is, however, a more general perspective from which this difference can appear artifactual. Both Grinnell and Elton describe niches as components of broader patterns, be they structures in physical environments or networks of functional interactions. If these patterns derive from or simply are causal relations, then both ecologists are giving causal representations, the only difference being the nature of the causal relata.[6] For Elton the relata would be functionally individuated sets of species (e.g., primary producers, herbivores, carnivores, guilds, and possibly individual species), which obviously depends on the resolution of the representation. For Grinnell the relations would be between more finely individuated biological units (species) and portions of the environment. This difference would then just reflect Grinnell's and Elton's different investigative priorities and explanatory commitments; the underlying content of the niche concept would be the same. The unified characterization would then be:

> **Niche**: a node in a nexus of causal interactions with abiotic and biotic factors occupied by a species.

In a fine-grained, species-specific way, "niche" would then simply convey causal information about ecological systems.

Despite the theoretical allure of unification, this proposal clearly fails. The problem is that Grinnell and Elton both countenanced the possibility of *empty* or *vacant* niches: unoccupied parts of environments (see Grinnell 1924, 227, quoted above) or biologically uninstantiated constituents of food webs (see Elton 1927, 27). The causally focused, species-specific characterization above seems unable to capture this broader notion; tracking the causal habits of nonexistent species is quite difficult. And this was not a superfluous conceptual aside. The idea was thought to do important work. Vacant niches feature prominently in Grinnell's explanations of the adaptiveness of species' phenotypes for specific environments, and

[5] Elton himself sharply distinguished his concept from Grinnell's. He criticized Odum's *Fundamentals of Ecology* for failing to distinguishing the two (Elton 1954).

[6] Suitably broadly construed, causal relations can represent both Elton's and Grinnell's target phenomena, but this general point will not be argued here.

Elton's explanations of putative ecological equivalents in distinct biological communities.[7]

One might think there is an easy fix, simply add "or not" to the characterization above. But the vacant disjunct is not so conceptually innocuous. Its addition seems to abandon the very causal information upon which the unified characterization is based. If it is the web of causal relations a species realizes that indicates the contours of the niche it occupies, no indication occurs in empty cases (from nonexistent species). Yet Grinnell and Elton were committed to the idea that niches endure rather than expire when those causal relations cease to exist (when niches are vacated). So they presumably cannot be (part of?) what individuates a niche. But what, then, is the relationship between the causal relations species participate in when they occupy a niche and whatever it is that characterizes that niche? Far from being "formalized by Grinnell" (Chase and Leibold 2003, 8), without an answer to this question the niche concept in its Grinnellian, Eltonian, or causal-unificatory guises is problematically vague.

Vagueness is not always problematic in science. Imprecision can accurately represent appropriate uncertainty about how a phenomenon is best described, or capture precisely the right level of generality when explaining it. It is problematic here because vagueness precludes clarity about what individuates niches. And clear individuation standards are certainly necessary if the concept is to perform the substantive function it is thought to in ecological theorizing. This kind of criticism has a well-established track record in biology. It is the same kind of charge made against adaptationism in an evolutionary context:

> The niche is a multidimensional description of all the relations entered into by an organism with the surrounding world.... To maintain that organisms adapt to the environment is to maintain that such ecological niches exist in the absence of organisms and that evolution consists in filling these empty and preexistent niches. But the external world can be divided up in an uncountable infinity of ways, so there is an uncountable infinity of conceivable ecological niches. Unless there is a preferred or correct way in which to partition the world, the idea of an ecological niche without an organism filling it loses all meaning. (Levins and Lewontin 1985, 68)

[7] A more recent example is Lawton's (1982) influential analysis of bracken herbivores in North American and Britain, and conclusion that the American communities were comparatively "unsaturated" – containing many more empty niches – and were therefore more susceptible to invasion.

Putting aside the allusion to uncountable infinities, and that the infinite divisibility of the world does not strictly entail that an infinity of niches is conceivable, it certainly does not seem that Grinnell's or Elton's accounts provide much guidance about how to affect the required partitioning.

The general problem is epistemological and evidentiary, not metaphysical and ontic. Vacant niches do not pose a problem because they violate some metaphysical criterion, such as that the causal approach would require niches be tokened (occupied). Ontologically, they *could* be construed as untokened components of causal type relations, uninstantiated dispositions, or in terms of counterfactuals with false antecedents. Knowing how to determine these types, dispositions, or counterfactuals is the difficulty. Without these theoretical possibilities being realized by occurrent species that would reveal such information, it is quite unclear how their nature can be ascertained.

For the kind of purpose Grinnell and Elton often seemed to have in mind when discussing niches, however, the inability to partition might be only marginally problematic, if at all. Rather than attempting to ecologically carve nature at some joint – one strictly inhabitable by a single species – they were often concerned with analogical inferences across dynamically and structurally similar communities. Evolutionarily distant communities in geographically remote areas sometimes seem to exhibit similar dynamics. If this African grassland community has a large species guild performing critical ecosystem function X, then that ostensibly analogous American grassland community exhibiting patterns quite similar to X might also have a guild performing that function. This type of inference is particularly clear in Elton's allusion to a basic "ground plan" underlying all animal communities. But the same idea plausibly underlies Grinnell's descriptions of how species are adapted to properties of their physical environments, and how at very disparate geographical locations one still sometimes finds similar kinds of species if their environments are similar. Lawton's (1982) bracken study is another example of the same kind of inference.

These analogical inferences depend on recognizing broad causal patterns that indicate similarities of structure or dynamics: the similarities make the analogies apt when they are. But analogy aptness does not require isomorphism, homomorphism, or any other fine-grained correspondence of dynamics or structure, the kind of close correspondence a fruitful niche concept that determinately partitioned the environment (Grinnell) or functional community relations (Elton) would arguably afford. Of course, analogical inferences are notoriously difficult to assess and highly defeasible. But when they are fruitful, it is characteristically not because such a high degree of correspondence precision can be established. Such precision in fact cuts against the less constrained connections analogies trade on. Biologically informed pattern recognition, not a foundational and systematic niche concept, seems to underpin the cross-community, cross-environment inferences Elton and Grinnell were making.

One might think the allusion to pattern recognition opens the door for a more positive evaluation of the niche. Judging that different biological systems exhibit similar patterns, it might be objected, is sufficient to ground the concept's utility, albeit in a vague form that matches the imprecision of "pattern." On, say, Elton's functional approach, considering the role played by species (or guilds, carnivores, or even more inclusive classes) would ground judgments of pattern similarity. On what other basis could such similarity be judged?

That biological communities do not seem to instantiate a universal or even widely generalizable "ground plan" at anything other than the most generic organizational level – and certainly not at the level of species – exposes one serious limitation of this claim. But the more basic problem is that it incorrectly reverses the direction of dependency. As noted earlier, Elton characterized niches in terms of this ground plan. Only via reference to that latter does the former acquire meaning; just as "vicar" acquires significance only in certain religious and institutional contexts.[8]

## 2 *The Struggle for Existence* and Competitive Exclusion

The shortcoming discussed above squares poorly with the prevalent view of the niche's role in perhaps the best candidate for a distinctively ecological law, the competitive exclusion principle (CEP). Simply put, it says *complete competitors cannot coexist* (Hardin 1960) or, in niche-theoretic terms, *species with identical niches cannot coexist*. This principle has a long history in ecology. Grinnell, for instance, arrived at an exclusionary principle early in the

[8] Unsurprisingly, the view of functions that perhaps best fits function talk in ecology, Cummins's causal role account (1975), exhibits the same priority. Only after a capacity of interest in a system being represented has been delimited can system parts be judged to have causal role functions. Talk of functions untethered from such a specification is simply confused on Cummins's account.

twentieth century, perhaps drawing on suggestive passages from Darwin (see Hardin 1960).[9] But the CEP's most compelling development and elevation to "principle" status is largely credited to Georgy Gause, his influential *The Struggle for Existence* in particular.

Despite its pedigree and pretensions to lawhood, the CEP is controversial. With Popperian flare, Peters (1991) deemed it tautologous. Even more methodologically tolerant ecologists have called it "untestable" and "of little scientific utility" (Pianka 2000, 248). The present task is not to render judgment on these claims (see Chapter 2 on ecological laws). Rather, it is to evaluate what could be considered the received view about the contribution the niche concept makes to CEP, which Griesemer (1992, 237) captures: "Gause's and Park's experiments showed that the concept of niche, in the guise of determinants of relations of competitive exclusion, was central to an understanding of population dynamics and the evolutionary structuring of communities." But if the concept is as problematically imprecise as indicated in Section 1, such centrality would be perplexing. No concept that indeterminate can do that much heavy theoretical lifting. Fortunately, a close reading of Gause's reasoning shows that although he used the term, the concept actually contributes little.

In several ingenious experiments, Gause (1934) studied competitive dynamics in paramecium and yeast species. In constant ecological conditions (e.g., nutrient levels, medium temperature, turbidity) and absent refugia that might mitigate interspecific competition effects, one species inevitably outcompeted the other to extinction. This result matches what classical Lotka–Volterra competition equations predict (exclusion), and Gause believed this furnished a compelling case for CEP. The key question is what the niche concept contributes to this case.

Gause (1934, 19) first mentioned "niche" in a context-setting discussion of "general principles" zoologists had developed in connection with competition. After citing Elton's (1927) "place in a community" definition referencing "habits, food, and mode of life," Gause then stated, "It is admitted that as a result of competition two similar species scarcely ever occupy similar niches, but displace each other in such a manner that each takes possession

[9] "[T]wo species of approximately the same food habits are not likely to remain long evenly balanced in numbers in the same region. One will crowd out the other" (Grinnell 1904, 377).

of certain peculiar kinds of food and modes of life in which it has an advantage over its competitor."[10] The clause "It is admitted" reflects Gause's awareness that Lotka, Volterra, and J. B. S. Haldane (1924) had already demonstrated exclusion with mathematical models of competition, models in which "niche" is absent (see below). The CEP was definitely "in the air" well before *The Struggle*.

But what is most striking about Gause's claim is how little Elton (1927) actually tied the niche concept to competition, and that he did not entertain anything resembling the CEP. In fact, Elton allowed that two species might occupy one niche (see Section 1). While many ecologists at the time were seeing competition as the prime driver of community structure (Kingsland 1995), Elton never shared this confidence. Elton was also quite skeptical of the ecological salience of mathematical approaches to studying natural systems; he thought they typically oversimplified their subject (Crowcroft 1991). Gause's effort to situate his project within the influential work of the day – *Animal Ecology* having had an immense impact on the incipient science – therefore seems to run a bit roughshod over the actual content of Elton's niche concept.

Immediately after mentioning Elton's niche, Gause (1934, 19–20) illustrated the idea of closely related species with different niches with an example of phylogenetically close sympatric tern species. They appeared to minimize competition by having distinct food sources. But the importance of differential feeding behaviors in communities composed of related species was well known long before Elton (or Grinnell) ecologically coined "niche," at least since Darwin's discussion of Galapagos finches. Moreover, modeling work bereft of niche considerations Gause knew well demonstrated the same result. Gause (1934) cited Lotka's (1932) analysis of competitive equations, which showed competitors could coexist by utilizing different food sources. The analysis never mentioned "niche," and Elton and Grinnell are not referenced. Without something beyond the mere fact that Elton's niche includes food, it therefore appears the concept contributes little to Gause's case for CEP in this part of *The Struggle*.

The aim of Gause's initial discussion was context-setting, however. And at the section's end Gause (1934, 19) emphasized, "we shall endeavor to express all these relations [food sources on competition] in a quantitative form."

[10] Gause (1934) did not cite and was apparently unaware of Grinnell's work.

His explanation twenty-five pages later of how Lotka–Volterra differential equations represent competitive dynamics supplies that quantification, and it contains the next "niche" reference. If the niche concept is to make a significant contribution to the CEP, this is the place.

At first glance, the intended contribution seems clear. "Niche" first occurs in this section in Gause's (1934, 45–46) discussion of $\alpha$, a coefficient of interspecific competition in the equations:

> This coefficient α shows the degree of influence of one species upon the unutilized opportunity for growth of another. In fact, if the interests of the different species do not clash and if in the microcosm they occupy places of a different type or different "niches" then the degree of influence of one species on the opportunity for growth of another, or the coefficient $\alpha$, will be equal to zero. But if the species lay claim to the very same "niche," and are more or less equivalent as concerns the utilization of the medium, then the coefficient $\alpha$ will approach unity.

Putting the potential significance of scare quotes aside, niches – via "places" – then seem to factor explicitly into Gause's (1934, 47) word-equation explanation of the equations given one page later (see Figure 2).[11]

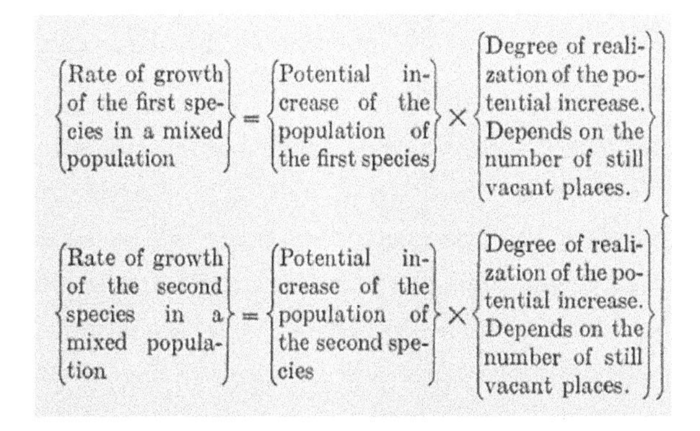


Figure 2 Gause's word equation representing standard Lotka–Volterra competition equations.

[11] It is worth noting that "places of a different type" clearly appears to be the intended synonym for "different niches." The mention of clashing interests seems to be a distinct consideration, one plausibly measured by the degree of competitive intensity between species. Whether the niche concept contributes to our understanding of this measure is, of course, the issue under evaluation.

Immediately after this verbal description, Gause (1934, 47) gave their mathematical representation:

$$\frac{dN_1}{dt} = b_1 N_1 \frac{K_1 - (N_1 + \alpha N_2)}{K_1} \tag{1a}$$

$$\frac{dN_2}{dt} = b_2 N_2 \frac{K_2 - (N_2 + \beta N_1)}{K_2} \tag{1b}$$

where $N_{1,2}$ represent competing species; $b_{1,2}$ represent birth rates; $K_{1,2}$ represent "maximally possible" carrying capacity population sizes; and $\alpha$, $\beta$ are competition coefficients representing the effect of $N_1$ individuals on $N_2$ individuals, and vice versa. These equations for the "struggle for existence," Gause (1934, 48) clarified, "express quantitatively the process of competition between two species for the possession of a certain common place in the microcosm." Note "place" here and in the word equation of Figure 2.

"Degree of realization of the potential increase" in the far-right term designates the "drag" factors represented in competition equations. Equation (1) shows there are two: intraspecific density-dependent drag captured by the logistic element $\left(\text{the second term: } \frac{-b_{1,2}N_{1,2}N_{1,2})}{K_{1,2}}\right)$ and interspecific density-dependent drag captured by the competition element $\left(\text{the third term: } \frac{-(\alpha,\beta)b_{1,2}N_{2,1}N_{1,2})}{K_{1,2}}\right)$. These factors, the word equation tells us, depend on the "number of still vacant places." What else could those places be but the same "places" Gause considered when discussing α and "niches" a single page before, or the "place" in Elton's niche characterization Gause referred to explicitly? The first chapter of *The Struggle* pays Darwin significant tribute for largely founding its scientific subject matter; perhaps Gause was harkening back to his niche-like use of "place" (see Pearce 2010).[12]

The terminological convergence here, however, is highly misleading. By "place" in "vacant places" and "common place in the microcosm" Gause meant something very specific, and entirely distinct from "place" in Elton's "place in a community." The former sense is made clear in Gause's earlier discussion of the logistic equation's representation of intraspecific density-dependence.

[12] In fact, adding to the terminological congruence, Gause ([1934] 2019) seemingly used it in precisely this sense on page 1: "Darwin considered the struggle for existence in a wide sense, including the competition of organisms for a possession of common places in nature, as well as their destruction of one another."

For species $N$ in a specific environment at a particular time, "The difference between the maximally possible and the already accumulated population ($K - N$), taken in a relative form, i.e., divided by the maximal population ($(K - N)/K$), shows the relative number of 'still vacant places'" (Gause 1934, 34–35). That is, "vacant places" simply numerically measure how much more a population can grow given intraspecific density-dependence and – when competing with another species – interspecific density-dependence. "Common places" are then actually arithmetic units of population size, realized or potential (vacant), that species "compete" for according to Gause.[13] Function-laden Eltonian notions of habit, feeding behavior, and mode of life are orthogonal to this numeric "place" notion. The former therefore does not help make the case for CEP via the latter.[14]

But this portion of *The Struggle* – the mathematically rich explanation of competition – is arguably its most compelling core: "Gause's great achievement was to give a clear exposition of the way that competitive exclusion, so often previously noted, actually worked" (Hutchinson 1978, 152). The ostensibly mechanistic detail of the account of competitive dynamics made it so convincing. If niche considerations contribute little if anything to that account, what real work is the concept doing in Gause's case for CEP?

The remaining five "niche" references in *The Struggle* bolster this judgment. They occur (with occasional scare quotes again) fifty pages later in Gause's discussion of several paramecium experiments that resulted in competitive exclusion. All those references appear in one passage questioning the result's relevance to real-world biological systems:

> However, there is in nature a great diversity of "niches" with different conditions, and in one niche the first competitor possessing advantages over the second will displace him, but in another niche with different conditions the advantages will belong to the second species which will completely displace the first. Therefore side by side in one community, but occupying somewhat different niches, two or more nearly related species ... will continue to live in a certain state of equilibrium. (Gause 1934, 98)

But nowhere did Gause explain how different niches must be to ensure coexistence, how niches could be individuated, or, most important, how

[13] This language is very strained. Cheetahs and lions compete for food, territory, and other resources. Saying they compete to "occupy" possible numerical population sizes is idiosyncratic at best.

[14] Note, moreover, that Gause never connects the earlier talk of "clashing species interests" with the term "niche."

niche considerations could help determine the competition coefficients required by the Lotka–Volterra competition equations. That different species typically utilize different food sources was well known well before Darwin, and existing models of competition had already captured the resulting interspecific dynamics with mathematical precision (e.g., Lotka 1932). Rather than constitute the indispensable conceptual core of Gause's work supporting CEP, his allusions to the niche seem more gloss than grist.

## 3 Hutchinson's *n*-Dimensional Hypervolume

For many ecologists, Hutchinson's definition in "Concluding Remarks" (1957) was a watershed moment:

> it was not until Hutchinson's "Concluding Remarks" that the niche concept was rigorously defined and its relationship to competition and species diversity rigorously explored . . . Hutchinson succeeded in combining both the Eltonian and Grinnellian concepts of niche into one model. (Real and Levin 1991, 180–181)

This "revolutionary" account (Schoener 1989; Chase and Liebold 2003) set the trajectory of niche-based theorizing in ecology for several decades.

Hutchinson's definition, which first appeared in a footnote of an earlier limnological paper (Hutchinson 1944),[15] characterizes two concepts, a species' *fundamental* and *realized niche*. For a specific species $S_1$, consider all the environmental variables $x_1, x_2, \ldots, x_n$ that affect $S_1$, which Hutchinson (1957, 416) emphasized includes both biological and (nonbiological) physical factors. If these variables are conceptualized as axes, they define an abstract $n$-dimensional space. The subset of this space in which $S_1$ can persist indefinitely (i.e., have positive fitness) is the fundamental niche of $S_1$, with an important qualification: persistence is assessed in the absence of all *competing* species. Not all other species are excluded in this assessment, as is sometimes incorrectly claimed, because some of the environmental variables that define the space are in fact species (e.g., $S_1$'s food resources, or obligate symbionts). Of course, species often do face competitors that constrict their range. The subset of the fundamental niche actually realized by $S_1$ given competitive dynamics is its realized niche.

[15] "The term niche (in Gause's sense, rather than Elton's) is here defined as the sum of all the environmental factors acting on the organism; the niche thus defined is a region of an $n$-dimensional hyper-space" (Hutchinson 1944, 20, n. 5).

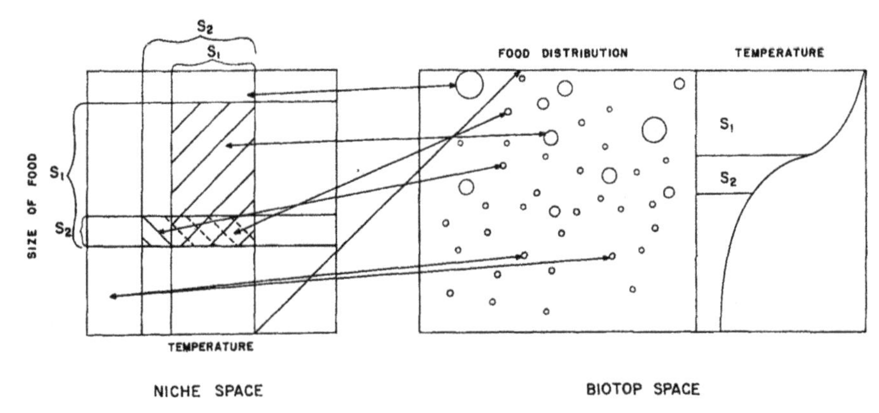


Figure 3 Hutchinson's diagram of two fundamental niches. From Hutchinson 1957, 421

In significant ways, Hutchinson's account breaks sharply with earlier work. Unlike Grinnell's niche but similar to Elton's, the fundamental and realized niches are abstractions, not portions of real-world environments. Any Hutchinsonian niche might correspond to highly disjoint sets of areas in the real world. The sole figure in "Concluding Remarks" illustrates this relationship, "biotop space" being the actual environment of the two species (see Figure 3).

Abstraction can make empirical concepts less tractable, and Elton's concept is often contrasted unfavorably with Grinnell's in this regard (Griesemer 1992). But Hutchinson's abstract definition is coupled with a significant conceptual shift that greatly enhances tractability: niches are strictly defined *in relation to species*, persistence of the latter determining the boundaries of the former (for fundamental niches). By definitionally tethering niches to species, Hutchinson's account is much clearer about how niches are individuated: positive fitness delimits the niche-relevant portion of a species' causal nexus. Concerns about how niches are delineable that plagued Grinnell's and Elton's conceptions do not gain nearly the same purchase. It may be exceedingly empirically difficult to ascertain, but in principle at least it is clear how species' niches can be delimited.[16]

[16] Note that what allows determination of niche boundaries is the focus on a particular species property affected by its environment, positive fitness. That focus makes it precise but also narrows the scope. If members of a species stray from their domain of nonnegative fitness and have significant impacts on other species or the abiotic environment, the source of those causal impacts would seem to fall outside the purview of Hutchinsonian niche considerations.

Gains in precision and tractability, however, came with significant costs. Hutchinson himself highlighted some shortcomings. For example, he claimed that the definition assumes all points comprising the hypervolume entail equal probability of persistence (1957, 417).[17] In reality, (absolute) fitness will vary markedly in any plausibly realistic niche space. Capturing these important differences requires explicit representation of the relevant ecological dynamics, that is, the more fine-grained causal details that determine whether and how species persist. The definition also assumes all the environmental variables characterizing the space can be linearly ordered; however, Hutchinson admitted, "In the present state of knowledge this is obviously not possible" (1957, 417). It is not entirely clear what precisely the difficulty is, and Hutchinson did not elaborate. Schoener (1989, 90) mentioned prey species and vegetation type as examples of non-linearly orderable environmental variables, without further explanation. It seems, however, that prey species can be ordered by their abundance, or frequency of encounter. If particular vegetation types are required for species persistence, and their existence is binary and not a matter of degree, then these environmental variables would not be linearly orderable. But such bivalence seems quite unrealistic. As habitats, patches of vegetation of different types presumably come in different degrees of suitability for different species. Suitability then seems to impose an ordering, from the optimally fitness-enhancing to the barely positive-fitness-maintaining. But this task does face the necessity of representing fine-grained causal details discussed immediately above.

Hutchinson also stipulated, but apparently did not perceive as problematic, that the environmental variables were independent and thus determined spaces with orthogonal axes. But this assumption is false in most cases. In the limnological systems for which Hutchinson first formulated the definition, for example, temperature, nutrient availability, light penetration, and other variables impacting species are all dependent on depth. Temperature and precipitation are interrelated for most if not all ecological systems. Nonindependence does not prevent construction of an abstract niche space,

[17] In fact, it is unclear whether his definition requires equality, and sometimes positive fitness (indefinite persistence) is fingered as the boundary of the fundamental niche (416). He also suggested that "Ordinarily there will however be an optimal part of the niche with markedly suboptimal conditions near the boundaries" (417) without any explanation of how his niche concept helps account for such variation.

but it necessitates nonorthogonal, skewed axes and coordinate systems to do so. Besides making visualization much more difficult, it also renders inapplicable some techniques used to analyze the detailed dynamics on which niche spaces depend (e.g., the Fourier method for partial differential equations representing those dynamics). And Hutchinson obviously could not have just stipulated that only independent variables delimit the (fundamental) niche on pain of losing information relevant to persistence. Nonindependent variables may nevertheless jointly bear on a species' persistence.

These are nontrivial problems, but they pale in comparison with the limitation imposed by the significant conceptual shift away from Grinnell's and Elton's account: defining niches *in terms of species persistence*. That relativization, for instance, makes the limited notion of a vacant niche Hutchinson's definition does afford much less explanatorily potent.[18] A niche absent an occupying species with the same explanatory potential as Grinnell's and Elton's is impossible because the former is definitionally dependent on the latter.

First, some clarity about what is possible on Hutchinson's definition. Consider when a species' realized niche is a proper subset of its fundamental niche, because of competitive exclusion as Hutchinson presumed, or other processes that prevent realization.[19] There is a clear sense in which this is an unoccupied niche on Hutchinson's definition: a species could have occurred here but does not. But notice how confidence about the modality is acquired. It derives from knowledge of species, in particular, what affects their persistence. *Given* this species and these persistence-relevant facts, *then* the contours of its niche can be identified. The direction of dependency is clear. But without a substantive, species-independent niche concept, Hutchinson's account is explanatorily impoverished. It does not have the resources to explain phenomena such as adaptive radiation into novel environments, degrees of "saturation" in ecosystem structure, similarities of different communities' dynamics, and others that were squarely in Grinnell's and Elton's purview. Explaining these phenomena requires showing how features of the physical environment, or perhaps a ubiquitous biotic ground plan, account for species doing what they do, or what they would do in certain circumstances.

[18] Hutchinson sometimes failed to recognize this implication of his definition (e.g., 1957, 424; 1959, 150; 1978, 161).

[19] Thanks to an anonymous reviewer for bringing this case to my attention.

Hutchinson's approach inverts this order. It therefore does not even seem able to account for ecological equivalents. If two species appear ecologically equivalent, then that view can obviously be *expressed* in terms of their Hutchinsonian niches. But identifying niches as objects of any comparative interest presupposes the prior judgment of equivalence; it cannot undergird it. The former therefore does not supply resources for *explaining* the latter. What could explain ecological equivalence is an understanding of the fine-grained causal details of what the species do, which, as stressed above, Hutchinson's account does not provide. In general, the idea that "the Grinnellian niche and the Eltonian niche are united through correspondence between points in N [the abstract niche space] and points in B [the biotop space]" (Real and Levin 1991, 181) in Hutchinson's definition just fails to recognize how dramatically his concept diverges from theirs, and how that divergence impacts its explanatory capabilities.

For the same reason, the Hutchinsonian niche cannot ground or otherwise be the basis of the competitive exclusion principle (CEP). As indicated above, competitive dynamics are excluded from the fundamental niche's characterization. The realized niche, on the other hand, *assumes* the principle: "we should expect that, in the part of the hyperspace where the overlap occurred, competitive exclusion would take place and the overlap would either be incorporated into the niche of one or the other species or be divided between the two, producing the *realized* or *postinteractive* niches of the two species" (Hutchinson 1978, 159).[20] Hutchinson, along with many

[20] There are technical complications in assessing the relevant "overlap" that expose further challenges of Hutchinson's hypervolume approach. First, if such overlap could be assessed in niche space, it would require determining the intersection of two subspaces in a more expansive space defined by the total set of environmental variables for both species. The fundamental niches of different species will almost always be defined with nonidentical sets of environmental variables, hence the need for the more expansive superspace. Although different species are sometimes similar in specific ways, which can generate competition, they almost always have significantly different ecological requirements and tolerances in other ways. But, second, and more important, it is unclear whether overlap can even be assessed in that abstract space. When competition occurs, it occurs in the real-world space species physically occupy (the "biotop" space). Trees compete for light and soil nutrients in geographically coincident portions of the rainforest; pelagic birds compete for nesting sites in specific cliffs of remote ocean islands. Competitive exclusion in the biotop space would then translate into exclusion in portions of the abstract (nonspatially explicit) niche space (see Figure 3) that do not intersect in any set-theoretic or geometric sense. Simply talking of overlap in an abstract hyperspace is therefore not an adequate representation of (spatially explicit) competitive dynamics.

ecologists at the time, thought competition was the primary driver of community structure; he called the CEP "a principle of fundamental importance" (1957, 417). Hutchinson's niche concepts reflect this commitment; they do not independently support it.[21]

Niches as defined by Hutchinson therefore convey very little information about community dynamics. Hutchinsonian niche considerations cannot, for example, determine which of two competitors will outcompete the other, and to what degree. And even delineating niches in the first place is empirically intractable when any more than a very small number of factors influence a species' persistence, which is rarely if ever not the case. In fact, answering most questions that ecologists find important – whether competitive exclusion will occur at all when fundamental niches overlap (without simply assuming it will), how it occurs, the dimension size $n$ of the hyperspace, the identity of those dimensions, and so on – requires an explicit representation of species dynamics. As a tool for representing and thereby understanding the dynamics responsible for community structure and species properties, the Hutchinsonian niche is hardly the epoch-making innovation it is often heralded to be.

## 4 Conclusion

These shortcomings of past accounts are not a historical curiosity. In a recent effort to reinvigorate niche-based ecological theorizing after Hubbell's (2001) influential neutral theory, Chase and Leibold review past characterizations of the concept and propose a definition aiming at synthesis:

> Niche Definition #1: the joint description of the environmental conditions that allow a species to satisfy its minimum requirements so that the birth rate of a local population is equal to or greater than its death rate along with the set of per capita effects of that species on these environmental conditions. (2003, 15)

The next sentence clarifies that the definition is "a simple joining of the two concepts that we have outlined in our historical review," by which they

[21] That commitment, furthermore, reflects but one view of what primarily governs the structure of biological communities. Fundamental and realized niches could be defined, for instance, with predation, mutualism, or other ecological interactions at the forefront. If competition is not the main driver of patterns and processes in the ecological world, Hutchinson's approach will miss much of what does.

mean a Grinnellian–Hutchinsonian concept and Eltonian one.[22] In effect, then, this is a bipartite notion:

> Niche = def (i) all factors causally relevant to a species' persistence;
> (ii) all the species' causal impacts on those factors.

But this characterization's sweeping generality raises serious concerns about its scientific utility. Rather than yielding something fruitful – that would, say, provide guidance in representing and analyzing community dynamics – this niche definition simply seems to acknowledge such dynamics exist. It is as if, in a chemistry context, one were told that the concept of "matter" – with no information about what specific forms it can take, its compositional building blocks, or its connections with other properties or lawful regularities – is the key to chemical analysis. What this niche definition offers seems largely to be a superfluous gloss on the causal details actually required to assess species persistence.

Chase and Leibold do not explicitly recognize this deficiency, of course, but they may suspect something is amiss about the first definition because they offer a second they claim is more precise:

> Niche Definition #2: the joint description of the zero net growth isocline (ZNGI) of an organism along with the impact vectors on that ZNGI in the multivariate space defined by the set of environmental factors that are present. (2003, 16)

"Zero net growth isocline" is short for the population values where $dN/dt = 0$. Revealingly, just before definition 2, Chase and Leibold (2003) say, "[W]e will use simple population dynamics models to justify a second more precise version of this definition [#1]." In other words, niches are only determinable, and this definition is only justifiable, once species interactions have already been represented in those models. That is, this niche concept contributes little or nothing to determining that representation.

In the rest of their book, this prioritization is never upended. It contains interesting extensions of resource utilization models first developed by Tilman (1982) and sophisticated analyses of how empirical data might bear

[22] That is inaccurate. Any partitioning of environments independent of species persistence considerations, Grinnell's conception, is absent. Species' functional properties that cannot be characterized relative to environments in any straightforward way, Elton's conception, are also absent. Rather, this seems to be a recasting of Hutchinson's notion.

on them. But one despairs of finding any nonredundant contribution the niche concept, as they define it, makes to these analyses. The absence is not surprising. Tilman's (1982) highly influential book mentions "niche" exactly four times, and in each case it refers to a label used by others.

The shortcomings described above do not impugn the sophisticated modeling and empirical studies done under a "niche" rubric. But they do indicate there is not some insightful and foundational concept at the base of this work, in some way guiding and shaping it all. That work stands alone.

# 2 Distinctively Ecological Laws and the Reality of Biological Communities

Apart from physics' long-standing exceptionalism – and probably chemistry after Mendeleev and others' triumphs – every special science struggles with the question of whether it possesses distinctive laws. Biology in general and ecology in particular are no different, and like scientists in many fields (e.g., economics, psychology) ecologists have partitioned pro and con (e.g., pro: Lawton 1999; con: Lockwood 2008). Most topics in the philosophy of science concern issues that involve a complex interplay of intricate conceptual and empirical considerations. This complexity is on full display in the debate about whether there are ecological laws. At the outset of the discussion two cautions should be heeded.

First, distinctively ecological laws may exist even if the best prospects from contemporary ecological science fall far short. What ecologists have thus far achieved, and whether there are distinctively ecological laws that future ecologists might (or might not) discover, are different issues. Different arguments made in the growing literature on "laws in ecology" often track one but not the other of these distinct issues.

Second, the qualifier "distinctive" makes this debate especially philosophically perilous. It stipulates that not just any law governing the entities ecologists study (organisms, populations, communities, habitats, ecosystems, etc.) will do. The candidate law must possess a characteristically ecological content; it must fall within the properly conceived purview of ecological science. Observing that plants and animals have positive mass and thus are governed by $E = mc^2$, or discerning that all creatures discovered thus far are carbon-based and therefore conform to certain lawful chemical regularities, do not fit the nomological bill. Note that this nomological shortcoming can persist even if these physical or chemical facts explain (perhaps even best or uniquely explain) ecological patterns and processes. For example, facts of geometry and underlying lawful chemical and physiological regularities might reductively account for many allometric scaling patterns ecologists have uncovered (see below). That would establish that laws govern entities in the ecological domain, but not necessarily that there are *distinctively ecological* laws. As we will see, gauging precisely what is "characteristically ecological" or within the "proper purview" of ecology is far from straightforward. But doing so is hardly a matter of abstruse philosophical trivia of negligible broader interest. Many take the possession of laws to be a signature mark of a science's maturity and objectivity,[1] so the stakes, both philosophical and scientific, are of the highest significance.

## 1 The Received View of Natural Laws

Some challenges to the idea that ecological laws exist can be dispensed with easily. One is the pervasive claim that ecological systems are too complex to admit of laws: they are simply too complicated, chaotic, and/or random to exhibit lawful regularities. Those adjectives, if applicable generally and unequivocally within ecology, would certainly help indicate the severity of the challenge of finding such laws. But irrespective of whether they do so apply, they would provide poor grounds for thinking ecological laws do not exist or could not eventually be found (Ginzburg and Colyvan 2004). It is difficult to imagine a more complicated system than the entire universe from the cosmological to the subatomic scale, but no competent scientist or philosopher suggests its complexity is not governed by relativistic and quantum mechanical laws, or that humans do not continue to uncover its underlying law-governed physical and chemical dynamics.

One might think the proposed similarity is misleading because the kind of mechanical and electromagnetic interactions at issue in theorizing in physics are much simpler than those in ecology. In ecology, there is a myriad of different kinds of entities and types of interactions of varying degrees of intensity between organisms, and between organisms and their abiotic environments (see Strong 1980). This charge shortchanges the hard-earned

[1] Part of what allowed biology to be a respectable target for philosophical analysis was the forceful rebutting of the charge, made most prominently by J. C. Smart (1959, 1963), that biology was not sufficiently systematic or rigorous to possess laws. It simply did not measure up to physics and chemistry. Once Ruse (1970) and others cleared the discipline's nomological credentials, philosophy of biology blossomed.

scientific clarity about the dynamics of the universe, clarity humans attained only after centuries of experimental and theoretical toil, and forgets the befuddling diversity of physical phenomena that confronted early scientists (and proto-scientists) at the beginning of secular inquiry. Only from a contemporary perspective do earthquakes, eclipses, water boiling and freezing, thunder and lightning, magnetism, and of course numerous other phenomena seem anything less than a bewildering multitude of disparate mysteries. Even the nature of inanimate matter once posed a seemingly insuperable challenge. Before Mendeleev, Lavoisier, and their scientific predecessors, the vast variety of substances in the world with distinct and strange properties must have appeared to defy non-supernatural explanation.[2] Complexity was a surmountable obstacle in chemistry and physics, and nothing about complexity in ecology appears to present, in principle, an obstacle to doing the same (but perhaps eventually).

Other challenges to there being laws in ecology are not as easily eluded. Critics have cited ecology's relative paucity of predictive success, that ecological models and experimental results too often lack sufficient generality or statistical validity, and that candidate laws are riddled with exceptions, to name a representative handful (Simberloff 2004; O'Hara 2005; Lockwood 2008). These are the canonical criteria against which lawhood is typically judged, and the criticism is that ecology fares quite poorly against them. Peters's (1991) famously caustic critique of ecology embraced an indefensibly strong falsificationism and claimed not only that evolution by natural selection is tautologous but that the same vacuity also infects much ecological theorizing. These striking claims rightly have few contemporary adherents, but significant portions of his critique stand independently and have some merit. In fact, the credible criticisms that many concepts in ecology are problematically imprecise, and that ecological theory falls markedly short of the epistemic virtues of other scientific theories that do trade in laws, has been amply voiced by philosophers with much less dubious theoretical commitments (e.g., Shrader-Frechette and McCoy 1993; Sarkar 1996; Cooper 2003) and by prominent ecologists themselves (McIntosh 1985; Lawton 1999; Simberloff 2004). The ultimate reason(s) for this failing remain

[2] It is therefore unsurprising that alchemy, steeped in mystical and theological supernatural appeals, was inextricably entwined with the origins of chemistry as a scientific discipline (see Bensaude-Vincent and Stengers 1996).

elusive, but perhaps complexity, though surmountable, has yet to be actually surmounted.

Some philosophers resist this conclusion. Those traditional markers of nomological status, they argue, are too stringent and should therefore be jettisoned in favor of a more relaxed conception. With the possible exception of quantum mechanics, or relativistic quantum field theory more precisely (Ismael 2015), none of the canonically labeled laws of physics or any other science satisfy such exacting standards (Ginzburg and Colyvan 2004). For example, general statements widely regarded as physical laws, such as ideal gas laws, Coulomb's law of electrostatics, and even Newton's laws, do not hold in many circumstances and thereby lack universality. Ecological generalizations commonly accorded nomic status fare similarly. For example, the putative latitudinal gradient in species diversity (see Rosenzweig 1992) and the so-called species–area power law (see Connor and McCoy 1979) are frequently considered law-like but nevertheless are known to have numerous exceptions. Only on a less stringent understanding of what natural laws are can the unparalleled epistemic deliverances of scientific inquiry attain the status they are typically accorded, and on that understanding ecology is a science that trades in laws.

One strategy for implementing this more forgiving view employs the idea of a ceteris paribus clause. A ceteris paribus clause is an exemption clause. It can accommodate possible exceptions and allow for limitations in the scope of putative laws, a clear departure from the traditional view that natural laws must be true and universally applicable, besides support counterfactuals and be empirical (as opposed to being logical or mathematical).[3]

[3] The traditional conception, due largely to the Scottish philosopher David Hume, has its recent detractors (see Hempel 1966, chapter 5, for a classical description of the traditional conception). For example, Sober (1997) argues that mathematical truths discovered in modeling biological systems – and thereby known a priori – should be deemed laws. Mathematical truths are certainly necessary in a way that contingent truths – such as the fact that five books currently sit on my desk – are not, and such truths do assist in describing and explaining biological facts when mathematical models are used to represent biological systems *and the representations are accurate*. Liberalizing the notion of natural law beyond the empirical domain might encompass these mathematical truths and thereby bestow them with the same status as recognized empirical laws from physics, chemistry, or elsewhere. But that possible philosophical payoff would come at a significant cost because it is thoroughly unclear what criterion could distinguish the purported nomological nonempirical truths from the nonnomological nonempirical ones. A nonempirical truth like "Bachelors are unmarried" is, of course, not nomological, but Sober (1997, S459) suggests that any "counterfactual-supporting, qualitative generalization, which describes how systems of specified type develop through time,"

For example, a suitably crafted ceteris paribus clause might insulate the nomological status of Newton's laws from the gravitational and relativistic contexts in which they fail to hold, even as useful "approximations." In the same way, such a clause might exempt the lawfulness of the latitudinal gradient in species diversity from the fact that a large set of organisms do not conform to the pattern (e.g., gymnosperms, mosses, parasitoid hymenopterans; see Kindlmann et al. 2007). Sober (1997) considers an explicit method of achieving this exempting functionality; fold the circumstances that ensure the truth of the putative generalization – call them the favorable initial conditions (*I*) – into the antecedent of a conditional (*L*) with the generalization as consequent (*G*):

$$L \quad I \rightarrow G.$$

Circumstances outside *I* incompatible with *G*'s truth, and hence that challenge its necessity, do not similarly challenge the conditional *L*. Perhaps *L* possesses a weaker form of necessity that is nevertheless sufficient for lawhood. This kind of conditionalization, and ceteris paribus clauses in general, might thereby supply a weaker notion of a natural law that many ecological generalizations would realize.

Even without fully entering the thicket of difficulties with ceteris paribus clauses (see Earman et al. 2002), there is a serious worry about the general inadequacy of this response. Absent clear and justifiable standards delimiting the extent of ceteris paribus clauses, this move appears ad hoc. No matter how unmotivated, without such standards nothing seems to forestall simply folding potential counterexamples into a clause's scope in a facile ploy to preserve nomic status. By this rationale, it seems that *any* true claim can be made faux-nomological with a well-chosen, sufficiently comprehensive ceteris paribus clause. The important insight that laws possess a kind of natural necessity that contingently true generalizations lack, even universally true generalizations such as "All bodies consisting of

deserves the label "natural law." The generalizations modelers in biology uncover satisfy the quoted clause, but the clause also seems to span the technical fruits of almost all mathematical modeling, no matter how far removed the mathematics may be from the dynamics of real-world systems. After all, any dynamical system of mathematical equations describes *some* type of system. Theorems concerning highly abstract differential or difference equations arrived at through esoteric proofs requiring no familiarity with actual biological or physical systems would then seem to count at natural laws. That price may be too high to pay for a permissive notion of natural law. See also Sober (2011).

pure gold have a mass of less than 100,000 kilograms" (Hempel 1966, 55), has little traction in this account.[4]

The shortcomings of this approach are compounded by the fact that most candidates for laws in ecology are based on models and theories that are *highly* idealized. That is, they incorporate unrealistic, that is, *false* assumptions about the systems they are intended to represent, largely to make model and theory analysis tractable. For example, they usually ignore some significant components and interactions of ecological systems (e.g., failing to include detritivores in standard community models), treat interactions as instantaneous and assume their effects propagate similarly (e.g., prey deaths contribute instantaneously to predator fecundity in standard predator–prey models), represent discrete components with continuous variables (e.g., almost all differential equations in the ecological modeling literature), describe community structure nonspatially (with the exception of individual-based models [see Chapter 4 on modeling in ecology], almost all community models), and so on. But these and other unrealistic assumptions warrant uncertainty about whether modeling results demonstrate properties of the real-world system a model is intended to represent or are byproducts of the idealizations themselves. And since it is often unclear what properties are primarily responsible for a system's dynamics given its complexity, particularly in ecology, idealizations may significantly mischaracterize their most important features. The enhanced sense of understanding conveyed by an idealized model may therefore fail to be about the system it is intended to represent, thereby misdirecting rather than assisting in the discovery of ecological laws. An additional concern is that technically inclined ecologists have often uncritically emulated mathematically sophisticated models of physics to ensure their modeling is similarly rigorous, but the emulation has sometimes led to serious misrepresentation of biological phenomena (see Justus 2008b). These difficulties are exacerbated by the short supply of

[4] Additional complexities confront this approach. For example, universality, or at least very broad scope, is usually considered a constitutive feature of laws. The fact that quantum mechanical (field) dynamics occur everywhere in the universe is a significant part of what warrants according the relevant physical laws their status. Scope is thus a reliable indicator of lawhood. What, then, is the scope of *L*? It is not simply *I*. If circumstances violate *I*, the conditional remains true. And if *I* ensures *G* obtains, then it seems the scope of *L* is universal. But it does not seem that such conditionals are universal in the same way quantum mechanical or relativistic principles are, or that they thereby accrue similar lawful credentials. How exactly the scope of these conditionals should be assessed and contribute (or not) to gauging their lawfulness is unclear.

extensive and long-term ecological data required to thoroughly vet model results, in contrast with the usually highly confirmed models found in physics and chemistry. Models are the main conduit through which theorizing occurs in ecology, so high degrees of idealization pose a significant impediment to finding distinctively ecological laws. Of course, the same concern holds for other areas of biology, such as population genetics.

Rather than see these aspects of scientific practice in ecology as an impediment, many believe they present an overdue opportunity to reevaluate and specifically weaken the notion of a natural law. Since a ceteris paribus clause insulates a generalization from counterexamples by restricting its scope and tying its truth to specific conditions, within the delimited domain and circumstances the qualified generalization could be taken to preserve the traditional conception. But perhaps even that suggestion is a philosophical vestige that should be abandoned: several philosophers have recently defended a much more accommodating view of laws, for biology in general (Elgin 2006; Mitchell 2009) and for ecology in particular (Cooper 2003; Mikkelson 2003; Ginzburg and Colyvan 2004). For example, Ginzburg and Colyvan (2004, chapter 2) argue that having exceptions poses no insuperable obstacle to ecological generalizations being lawful because many reputed laws of physics are also riddled with exceptions. Strictly speaking, disturbing forces from other masses prevent any planet's orbit from being an exact ellipse as Kepler's laws require; despite the example's centrality in physics textbooks, all billiard ball collisions violate the law of the conservation of momentum given the heat the collisions generate; and even Newton's mechanics breaks down at velocities approaching the speed of light, at very small scales, and for very large masses and associated gravitational fields. If the "law" rubric is appropriate in physics, why would it not be for similarly qualified generalizations in ecology?

These labels are obviously quite common and this line of reasoning trades on their legitimacy. But linguistic practice often poorly guides philosophical and scientific judgment. In numerous cases claims called laws seem decidedly mislabeled. For example, Hooke's "law," Mendel's "laws," Ohm's "law," and many others retain the label in scientific parlance, but findings after their discovery and promotion to lawhood have revealed they are actually false, and useful only to a degree as approximations within a specific and often quite narrow domain. It may be true that Newton's laws are all humans need to get to the moon and back (along with some fluid dynamics, metallurgy, propellant chemistry, mammal physiology, among many others), but what Newtonian mechanics says about the nature of space and time, quantities invariant across reference frames, and simultaneity is in fact incorrect. In part, calling a discovery a law honors its discoverer. But even the most exalted achievements of scientific inquiry may eventually be revealed to fall short of the epistemic credentials that appeared to merit the original judgment. If linguistic practice in physics exhibits this shortcoming, it should not be repeated in ecology.

There is a second worry about such a permissive notion of natural law: it seems to preclude distinguishing laws from mere regularities. If laws can have exceptions and need not be true or universal, how *exactly* can they be identified among the multitudes of well-confirmed generalizations scientists have uncovered and continue to uncover? And if laws are merely very well-confirmed generalizations with an impressively wide scope, perhaps also with a certain preeminent cachet, is there no difference *in kind* between laws and particular facts? After all, even the happenstance that only "silver" coins currently reside in my pockets supports various counterfactuals (e.g., "if I had purchased a single penny candy with the coins in my pocket, I would have received change") as well as various predictions (e.g., "if I pull a single coin from my pocket, it will either be a nickel, dime, quarter, half-dollar, or 'silver' dollar"). Perhaps there are no differences in kind within the set of empirical truths, merely differences in degree along dimensions of confirmation, explanatory power, scope, and so on, as many have recently argued. But that seems to be as much an argument that there are no laws as an argument for a new conception of what a law is.

## 2 Scaling Patterns in Ecology

One class of ecological generalizations avoids many of these model-based difficulties, and each generalization is widely regarded as very well confirmed, albeit with significant "scatter" of data around the proposed relationships. These are the so-called macroecological allometries (see Ginzburg and Colyvan 2004, chapter 2; Marquet et al. 2005). They include:

1. Kleiber allometry: basal metabolic rate is directly proportional ($\propto$) to a 3/4 power of body mass, that is, $(\text{body mass})^{3/4}$. First noticed by biologist Max Kleiber (1932), the larger the organism, the greater (at a 3/4 power) its calorie consumption rate at rest.

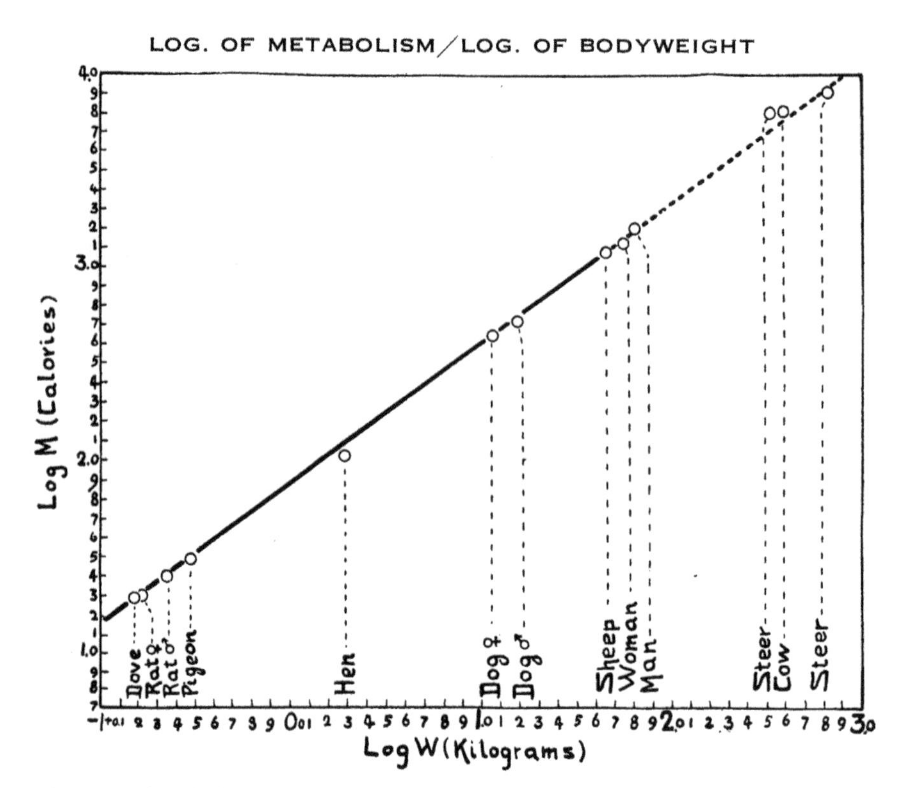


Figure 4 The Kleiber allometry between basal metabolic rate and body mass. From Kleiber 1932, 321

2. Generation-time allometry: organismal maturation time $\propto$ (body mass)$^{1/4}$.
3. Fenchel allometry: maximum reproduction rate is inversely proportional to (body mass)$^{1/4}$; first studied by Tom Fenchel.[5]

Other ecological allometries exist, but these are perhaps the most empirically vetted and thus strongest candidates for lawhood. Besides their high degree of confirmation, their broad scope (across *all* organisms) also seems to evince nomological credentials. The Kleiber allometry, for example, has been verified for organisms with masses ranging from elephants to bacteria (see Figure 4). For these reasons, Ginzburg and Colyvan (2004) suggest these allometries deserve to be called laws.

[5] Ginzburg and Colyvan (2004) state that the Kleiber allometry is the most empirically well supported and that the generation time and Fenchel allometry are likely based on it. For a critical perspective on the generality and evidential basis of the Kleiber allometry, see Dodds et al. (2001) and Glazier (2006).

There is, however, a formidable challenge to this view: John Beatty's (1995) evolutionary contingency thesis (ECT) that all distinctively biological generalizations describe, and perhaps can only describe, contingent outcomes of evolutionary processes.[6] The thesis presents a two-horned dilemma for candidate biological laws. If they are distinctively biological, then Beatty argues they are contingent because the evolutionary processes responsible for their existence and their "distinctively biological" status are highly contingent. The unguided, largely "blind" nature of genetic mutation, the similarly random nature of evolutionary drift, and the fact that natural selection acts with respect to environments that frequently (and contingently) fluctuate are prime examples of such contingency. This contingency, Beatty argues, is incompatible with lawhood in the same way that, for instance, the contingency of the universal truth "There are no 50 $m^3$ gold spheres," is incompatible with lawhood. If Bill Gates, Carlos Slim Helú, or perhaps a powerful alien with gold fever desired such a lustrous orb, nothing in the fabric of the cosmos would preclude its creation. Similarly, for any particular biological truth, physical and/or chemical circumstances could just as easily have conspired to falsify it (a more [or less] massive sun, an intragalactic gamma-ray burst, a colossal meteor strike, a less hydrophilic chemical composition of the earth, absence of the meteor strike that ultimately eradicated the dinosaurs, etc.). The very evolutionary processes that have shaped the biological world into its present state are nonetheless too weak to undergird the necessity lawhood requires.

Beatty's view is not without its detractors (e.g., Sober 1997), but that controversy need not detain us here because regardless of whether the allometries succumb to the dilemma's first horn, they do seem to fall to the second: they do not seem distinctively *ecological*. Take Kleiber's allometry. One compelling account of the pattern is that it is a byproduct of *none*cological factors: fluid dynamics and the geometric structure of circulatory, respiratory, and vascular systems of animals and plants (West et al. 1997). The ultimate explanation of that geometric structure, moreover, is also unlikely to depend on ecological principles or particular ecological facts. As such, whether or not this allometric fact should be accorded nomic status, its standing as a distinctively *ecological* generalization is questionable. If the

[6] For more exploration of the contingencies at issue, see Beatty 2006. Turner and Havstad (2019, section 6), give a very helpful overview.

pattern just follows from geometry and physics, why would it be any more distinctively ecological than the fact that all organisms have mass, or that organism body mass tends to scale with its volume? These are certainly truths concerning entities studied in ecology, but ecological science seems to contribute nothing nomological to their status.

This example broaches a general issue about the relationship between the nomological distinctiveness in question and reduction (see Chapter 4 on modeling in ecology). If West et al.'s (1997) analysis is correct, Kleiber's allometry reduces to facts of geometry and physics. The same kind of reductive relationship seems to hold for claims such as that all organisms have mass or that the effects of organismal interactions propagate at speeds less than the speed of light ($c$), for all organisms $E = mc^2$. The reduction in these cases is admittedly uninteresting: specific (and trivial) facts that organisms have mass and their interactions are decidedly subluminary combine with universal physical laws to produce lawful, albeit uninteresting statements. But ecological considerations contribute nothing to the statements' lawfulness, and that seems to dictate against them saying anything characteristically ecological. The same reasoning challenges Kleiber's allometry. If what determines the scaling pattern – what ensures its truth – are facts outside ecology, then that casts doubt on its distinctively ecological status as a candidate law. In general, if a lawful generalization X studied in one science reduces to facts in other sciences, that reductive relationship bears on the character of X, in particular, whether it should count as a distinctive law of the first science, its reducing sciences, or neither. This line of reasoning obviously has broad implications if reductive relationships hold widely between different sciences and the phenomena they study. For example, if all sciences ultimately reduce to physics, perhaps no other science has distinctive laws. Ecology would then just be one of the numerous nonautonomous "higher-level" sciences that lack this credential.

On the other hand, the uncertainties surrounding the status of the Kleiber allometry might cut in a different, more radical direction. Prima facie, the meaning of the modifier "distinctive" in the evolutionary contingency thesis seems rather straightforward. After all, different sciences are housed in different departments and given distinct labels for a reason, largely because they address different subject matters. By connoting a demarcateable content, a constitutive character of a science that defines its nature, the proposed notion of distinctiveness would simply reflect customary disciplinary divisions. But interpreted too strongly, the notion loses credibility. Sciences actually do not partition neatly, and extant academic boundaries do not appear to originate from some defining essence, but rather stem from historical contingencies and often extraneous socioeconomic factors. What, then, is the weaker yet defensible sense of "distinctive"? In the original essay, Beatty (1995) never explicitly defines the adjective, and the implicit indications of its content are indefinite; whether Kleiber's allometry should count as a distinctively biological generalization, for instance, is unclear.[7] Groping for a sufficiently clear notion, Sober (1997, S460) suggests that the "distinctive vocabulary" of a generalization determines its distinctive status. But that criterion is too permissive: theorems derived from nonempirical mathematical models of interesting but nonetheless nonexistent biological systems would arguably be distinctively biological by this measure, as it appears would the trivial reductive examples from above (e.g., "All haploid invertebrates have positive mass" certainly employs distinctive biological vocabulary). What these worries make clear is that the relevant notion of distinctiveness is quite unclear. If Sober (1997, S461) is right that "[t]here probably is no point in disputing how the phrase 'distinctively biological' should be understood," as our discussion seems to suggest, then the entire question of whether there are laws in ecology, or any other special science, might be poorly formulated.

## 3 The Competitive Exclusion Principle

If the litmus test could be "I know it when I see it," there clearly is a famous and distinctively ecological generalization with pretensions to lawhood, the competitive exclusion principle (CEP). Although Grinnell (1917) drew on Darwin's work to arrive at the same kind of exclusionary principle a few decades before and Darwin himself entertained a similar idea (see Hardin 1960), CEP's origination is largely credited to the Russian biologist Georgy

[7] In the light of the above discussion, are the Kleiber and other allometries "just mathematical, physical, or chemical generalizations (or deductive consequences of mathematical, physical, or chemical generalizations plus initial conditions)" and hence not biologically distinctive (Beatty 1995, 46)? From what Beatty says there, it is difficult to adjudicate this question (particularly given the qualifier "just"). Notice one important implication of this portion of the evolutionary contingency thesis: if Laplacian determinism is true (see Earman 1986), there are no distinctively biological generalizations.

Gause. In a series of brilliant experiments, Gause (1934) studied competitive dynamics in paramecium and yeast species. In constant ecological conditions (e.g., nutrient levels, water temperature, turbidity, etc.) and in the absence of refugia that would mitigate the effects of interspecific competition, one species inevitably outcompeted the other to extinction. Remarkably, this experimental result matched the exclusionary outcomes the standard Lotka–Volterra competition equations predicted. Given the (rare) congruence of empirical and theoretical results, Gause generalized the CEP: species with identical niches, that is, two species that would compete for exactly the same resources, cannot coexist (although see chapter 2 for a discussion of the niche concept's role here). The intuitive appeal of the idea and its apparently exceptionless status across many different biological systems has prompted its honorific designation as an ecological law, Gause's law.

Apart from the extreme case of exclusion, the degrees and types of "niche overlap" that permit coexistence became an important focus of contemporary attempts to explain species distribution patterns and dynamics in biological communities (Abrams 1983). The ecologist Robert MacArthur (1958) was one of the first to rigorously document this kind of phenomenon in his rightfully famous dissertation work on warblers in New England. The objective of the study was to determine how so many behaviorally and physiologically similar bird species could coexist in boreal forests, which seemed to contradict the exclusion principle. With such similar properties, it seemed that interspecific competition would be especially strong between the birds and would eventually lead to the extirpation of all but one competitively dominant species. Through meticulous observation MacArthur uncovered the mechanism that eluded the exclusionary outcome: different species bred and fed in distinct spatial parts of coniferous trees, and, furthermore, warblers exhibited strong territoriality toward those parts (Figure 5).

He also found that nesting times differed across the warbler species. This affected the same minimization of competition as spatial partitioning. These behaviors effectively divided the homogeneous arboreal habitat into disparate sections, thereby partitioning (spatially and temporally) the putative niche space. This process curtails competition and allows the extant set of warblers to coexist.

Despite these successes, there are reasons to doubt that the CEP constitutes a natural law. Perhaps the most acute worry is the concept of a niche itself. As Chapter 1 made clear, the niche concept is problematically imprecise. Any candidate ecological law trading in that concept inherits those

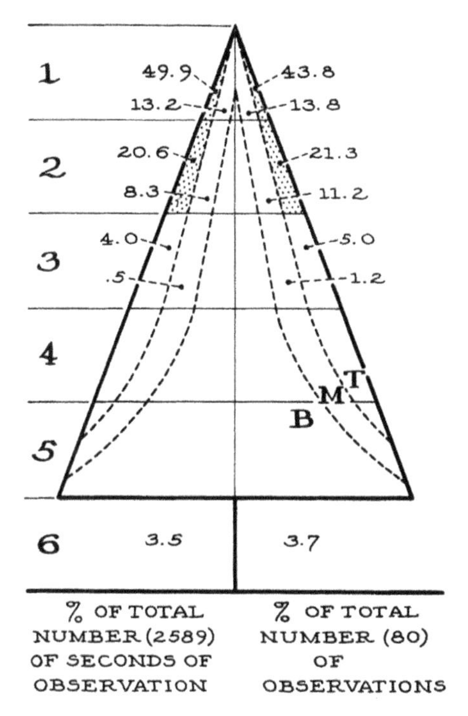


Figure 5 Representation of the arboreal feeding locations of the Cape May warbler. From MacArthur 1958, 602

deficiencies. But even if these conceptual inadequacies can be surmounted (or circumvented), there are further difficulties specific to the CEP. First, as a generalization the CEP is of quite limited scope. For example, it is inapplicable when resources are abundant and species are not competing for them. Nor does it apply in fluctuating environments where niche contours are ephemeral, or when changes in the direction of competition occur before exclusion can catalyze extinction. Since most environments nontrivially fluctuate, this is a serious limitation, and it recalls difficulties with ceteris paribus clauses described above. CEP also seems to have numerous potential counterexamples. For example, migration into an area can prevent the predicted competitive exclusion. Another apparently recalcitrant counterexample was originally identified by MacArthur's advisor, G. E. Hutchinson (1961): the seemingly inordinate number of plankton species given their seemingly simple, homogeneous environment. Known as the "paradox of the plankton," this issue remains an active area of contemporary ecological research and is yet to be conclusively resolved (see Tilman et al. 1982).

There is another threat to CEP as law: its empirical status. To appreciate the potential difficulty, first note that the relevant niche concept CEP invokes must be the Grinnellian, environmentally based notion (see chapter 2). In his seminal publication, Gause (1934) actually appeals to Elton's (1927) work, but Elton's functional niche concept would make CEP false. Different species often serve veritably identical roles in the causal nexus of interspecific interactions comprising a biological community. Different African grazers that migrate and ruminate together and different pollinators are but two plausible examples. In fact, ecological literature has labeled species with similar resource requirements that utilize them in similar ways a "guild" to capture this commonality. Instead, what permits species with similar causal roles in a biological community to coexist, according to CEP, is their partitioning of the environment. For example, two fish-eating waterfowl, the great cormorant (*Phalacrocorax carbo*) and the European shag (*P. aristotelis*), exhibit approximately the same causal interactions with other terrestrial species, but they nest in different portions of cliffs (as well as consume fish from different sources: estuaries and harbors vs. the open sea, respectively) (Lack 1954).

But with "niche" construed environmentally, the empirical content of CEP has been questioned. The problem begins with imprecision. For example, the CEP says nothing about what precise degree of niche differentiation is required to ensure coexistence. It seems that this can be answered only on a case-by-case, ecosystem-by-ecosystem basis, if at all. The poor general guidance sets up a troubling scenario. The reasoning proceeds as follows. Suppose two species coexist but ecologically appear very similar. Then, by CEP, their niches must differ. Their ecological similarity initially suggested similar niches, so at what point does investigation of how the species utilize resources and interact with the environment that reveals no significant difference constitute a counterexample to CEP? Without precise guidance about what degree of niche differentiation coexistence requires, the worry is that the CEP is effectively immune from empirical challenge. For this reason, Pianka (2000, 248) calls CEP an untestable hypothesis "of little scientific utility." The claim about utility should be rejected. As Slobodkin (1980) convincingly argued and the history of the science confirms, CEP has played a very useful role in ecological theorizing as a research heuristic. But scientific utility should not be confused for nomic status. The mechanistic worldview was extremely valuable in the development of science, but it was neither a law nor true (see Dijksterhuis 1961).

## 4 Are Biological Communities Real?

At a minimum, a biological community is a set of populations of different species. Usually, the species are taken to interact in *some* way to *some* degree. Beyond these platitudes, controversy emerges. The difficult question is whether communities are "something more" than the individual organisms of different species comprising them. If they are not, presumably they possess no independent existence. If they are, an account is needed of: (1) this "something more" *and* (2) how it confers independent existence. Absent either, realist aspirations are frustrated.

Assessments of item (1) primarily focus on the nature and intensity of interactions between species comprising candidate communities. This requires a careful dissection of the causal structure of these interactions, their dynamics, species distribution patterns, and what they reveal about how groups of species might be assembled into communities. Appreciating that this is a conceptual *and empirical* issue is essential. Assessing item (2) involves delving into the murky depths of metaphysics, principally to determine whether the additional causal structure these ecological assemblages possess actually "cuts nature at its joints," a proverbial criterion for ontological credibility according to most scientific realists. Ecological science seems to offer little or no new insights regarding the ontological question of whether causal novelty confers independent existence at issue in item (2), and most philosophical analysis has thus far concerned item (1).

Different positions on item (1) fall on a spectrum. At one extreme is the view that communities are simply aggregations of species at a particular location and time and that their relationships with the abiotic environment, not other species, largely determines their co-occurrence. On this view, communities as distinct ecological units are no more real than a *collection* of knick-knacks on a mantel, as opposed to the knick-knacks themselves, is real. The contingency of co-occurrence and lack of significant interaction are considered marks of the unreal. At the other extreme is the view that communities are tightly causally integrated units that exhibit a degree of functional cohesion, similar to individual organisms. On this view, communities are as real as individual organisms that possess these attributes.

A long-standing debate about the mechanism(s) of ecological succession in early twentieth-century ecology helped catalyze this question about the nature of biological communities. On one side were "holists" such as

Fredric Clements with his climax account of succession. For Clements, the specific constellation of abiotic factors in a given locale – cloud cover, elevation, precipitation, soil type, temperature, and so on – yields a deterministic sequence of succession stages that usually culminates, if undisturbed, in a final climax community. Different constellations usually produce different sequences and climax communities: grassland, mangrove, marsh, pine forest, and so on. If the seemingly mechanistic determinism were not controversial enough, Clements added that climax communities were "superorganisms" with the teleology and functional integration among constituent species that that term suggests. On the other side of the debate were "individualists" such as H. A. Gleason and H. G. Andrewartha, who discerned much less structure to succession, less cohesion in reputed biological communities, and generally more contingency underlying ecological patterns.[8]

The extreme positions associated with this debate have largely been abandoned as implausible, but the theoretical contrasts it drew continue to reverberate throughout contemporary ecology. For example, although both sides of the dispute considered themselves respectable empiricists, one influential criticism of the climax, "superorganism" theory was that it departed significantly and unjustifiably from what observations of succession, species distributions, and dynamics in supposed biological communities actually indicated. Robert Whittaker made this kind of criticism based on very influential analyses of plant distributions along environmental gradients, such as elevation. Plant species distributions along environmental gradients seem to overlap continuously and significantly, and do not form discrete identifiable boundaries (see Whittaker 1956). But, the argument goes, communities are only real if they have such distinct boundaries. So they are not real.

The key claim being challenged is that species distributions display converging boundaries that communities must possess. Odenbaugh (2007) gave three forceful responses to this argument. First, Odenbaugh rightly points out that Whittaker's results are inflated in claims such as: "Whittaker found, each species behaved totally independently . . . there is no such thing, really, as a pine forest, or a mixed hardwood forest or a tall-grass prairie or a tundra" (Budiansky 1995, 86) and "There are no discrete communities of

[8] See Eliot (2011a) for a historically engaging and conceptually rich account of this debate, one that locates the divergence more in methodological disagreements about how ecological research should be conducted than contrasting ontological commitments.

plants. The reality is endless blending" (Colinvaux 1979, 72). These audacious claims simply assume Whittaker's results can be extrapolated to all plants. But it is worth noting that both of these assertions occur not in top-tier scientific journals but in works of popular science where epistemically unwarranted flights of rhetorical fancy are less constrained. More measured assessment suggests Whittaker's results constitute strong *evidence* against the existence of the needed boundaries.

Odenbaugh's additional responses address this more reasonable interpretation. He claims that two implicit assumptions seem to underlie Whittaker's analysis:

> Interactions among species should be similar at *all* points along environmental continua. Thus, if two or more species interact in a certain way at a point, then if they interact at other points it is in the same way.... Groups of species should be associated at *all* points on a gradient if interdependence is to be accepted. Thus, if two or more species interact at a point on a gradient, then they interact at *all* points on that gradient. (2007, 635; emphasis added)

Odenbaugh correctly notes that these assumptions are false, but it seems implausible that Whittaker or other respected ecologists would be foolish enough to endorse such categorical claims. Exceptionless patterns are a rarity in ecology, and biology generally. One should no more accept the conditional that species interact at *all* points on an environmental gradient if they interact at one, than one should accept that species do not interact at all points if they fail to at one. The same goes for the functional form of the interaction. The degree of continuity of species interactions across environmental gradients is obviously an empirical issue. But all that is required for Whittaker's results to constitute evidence of a lack of boundaries is that interactions persist and remain similar along sufficient portions of the environmental gradient sampled. What counts as "sufficient" depends on the patterns Whittaker found and the species involved: the higher the turnover in species distributions, the smaller the portion generally required for sufficiency. Intraspecific variability exists, but organisms comprise a species in large part because they share a very similar ecological profile. One would expect, although only data and sophisticated statistical analyses could reliably confirm, that this similarity would ensure the sufficiency required.

Odenbaugh's last response is that Whittaker himself employed a "community-level" property, niche differentiation between two species, to explain the absence of boundaries in species distributions. This, Odenbaugh suggests, shows that Whittaker was himself committed to the existence of communities. If Whittaker's contention was to show that no communities exist and two species interacting shows they do, Whittaker would have indeed blundered. But the worry is that this characterization of the debate construes the goalposts too weakly. Surely what is at stake with respect to the reality of biological communities – and the target of Whittaker's analysis – is more than whether there are binary community-level properties such as predator–prey interactions between populations, and thus two-species-member "communities." If this were the issue, simply observing a population of lions consuming gazelles or parasitic mistletoes infesting a deciduous forest would be sufficient to resolve the debate. Even if they stop short of endorsing full Clementsian "superorganismal" status, proponents of the reality of biological communities have a more ambitious agenda. They believe there are many communities composed of many more than two species, that maintain some type of homeostasis, and that exhibit other kinds of causal integration that two-species interactions, which are often highly unstable, do not. Odenbaugh's criticism therefore seems to invoke an indefensibly weak burden of proof given what is at issue in this debate.

Against this more exacting standard, Sterelny (2001) gives a positive argument for the reality of communities based on paleoecological data and an analogy with organisms and species. What makes the latter real, according to Sterelny, is their internal regulation. For example, organisms maintain a boundary between themselves and the external environment, and they regulate their internal states against environmental changes. In particular, organisms exclude foreign objects and agents of disease through a complex array of processes. Species also regulate membership through behavioral, physiological, and genetic impediments to reproduction. Through an impressive regimentation of paleoecological data and scientific analyses to understand that data, Sterelny convincingly argues that there have been episodes of "coordinated stasis":

> Suites of species, drawn from quite different lineages, appear together quite suddenly in the fossil record. They persist together largely intact. The periods of persistence are evolutionarily significant: often a few million years. These species not only persist together; they do so maintaining both their morphological and their ecological characteristics. The commonest species stay common; the relatively rare stay relatively rare. Few new species migrate

> in, or evolve in place. Few of those in place at the establishment of an association disappear before the association breaks up. In general, *associations* persist, not just the individual taxa that make up those associations. Each seems to end with the association dissolving and many of its component species disappearing, to be replaced by a different but persisting assemblage. So the pattern is one of both evolutionary and ecological stability bounded on each side by a turnover event. (438–439)

Most of Sterelny's analysis is concerned with scrutinizing proposed explanations of these periods of surprising constancy based on the best theories of community dynamics. None proves successful. It therefore remains a tantalizing prize for future theorizing. It is abundantly clear, however, that strictly individualist approaches such as those associated with Whittaker uncontrovertibly fail to account for this phenomenon. The paleoecological data Sterelny considers therefore constitute strong evidence for the community-level internal regulation that, by analogy with the organismal and species cases, indicates the reality of biological communities. Sterelny does not over-extrapolate the significance of these results, and other paleoecological studies seem to provide countervailing evidence. For example, Davis and Shaw (2001) found that after the glaciers receded, individual tree species dispersed at different rates, in different directions, and from different origins. This seems to confirm the individualist view in which abiotic factors, not biotic interactions, drive species distribution patterns. As with most philosophical questions in ecology with an empirical component, the evidentiary issues involved are far from resolved.

# 3 The Balance of Nature

The commitment to some type of balance was a staple of the schools of natural philosophy from which biology emerged, long before the term "ecology" was even coined (Egerton 1973). For example, Darwin, with most of his scientific contemporaries, was committed to the idea of such a balance:

> Battle within battle must ever be recurring with varying success; and yet in the long-run the forces are so nicely balanced, that the face of nature remains uniform for long periods of time, though assuredly the merest trifle would often give the victory to one organic being over another. (Darwin 1859, 73)

Darwin and other early ecologists continued this tradition by attempting to derive the existence of a "natural balance" in biological populations from organismic metaphors and analogies with physical systems, although the analogical and metaphorical content often differed (see Kingsland 1995). For example, the ecologist Frederic Clements (1916) is best known for claiming to find functional integration within biological communities that resembled the physiological integration within individual organisms, and which justified conceptualizing communities as a kind of superorganism with analogous homeostatic properties. But Darwin (1859, 115–116) employed the same metaphor several decades before, with a much less problematic aim:

> The advantage of diversification in the inhabitants of the same region is, in fact, the same as that of the physiological division of labor in the organs of the same individual body.... No physiologist doubts that a stomach by being adapted to digest vegetable matter alone, or flesh alone, draws more nutriment from these substances. So in the general economy of any land, the

> more widely and perfectly the animals and plants diversified for different habits of life, so will a greater number of individuals be capable of their supporting themselves. A set of animals, with their organization but little diversified, could hardly compete with a set more perfectly diversified in structure.

Although this conclusion plausibly holds for communities in relatively constant environments and thereby provides a plausible explanation of the greater species diversity found in the tropics than in more environmentally turbulent temperate regions (see Rosenzweig 1992), later ecologists would show that specialization often constitutes a handicap in fluctuating environments that favor adaptable generalists (e.g. Pianka 2000, chapter 8).

There were two threads to Darwin's view on the character of this putative balance, particularly the causal forces responsible for it. Most scientists before Darwin did not fully appreciate the extent to which inter- *and* intraspecific competition shaped communities (Bowler 1976). A balance of nature was considered the result of a predetermined harmony that competition would only undermine. Darwin's balance was undergirded by a much more realistic dynamics. Interspecific competition constrains the populations comprising biological communities by limiting organisms' access to the resources they need to metabolize and ultimately reproduce. This curtails populations' geometric tendency to increase. Other forms of interspecific interaction have similar consequences. Predators and parasites, for instance, inhibit prey and host populations. Intraspecific competition produces the same inhibitory effect within a species, and it can inhibit other species through interspecific relationships. For example, intraspecific competition among prey limits predator populations.

But, as Darwin was well aware, these inhibitory relationships do not alone account for the kind of dynamic balance ostensibly exhibited in the natural world. The problem was the differential power and scope of intra- and interspecific competition. Intraspecific competition is fully general: it arguably occurs in all biological populations (but see Cooper 2003, chapter 3). But its power to restrain population growth is governed by the availability of resources. When resources are plentiful, little check on growth occurs. On the other hand, interspecific competition (predation, parasitism, etc.) can suppress population growth more effectively than intraspecific dynamics in such cases, but it is not universal: not all species seem to be connected in inhibitory interspecific relations. Thus, although intraspecific competition would limit all populations when resources were scarce and interspecific interactions would sometimes suppress growth further, if these were the only checks on populations, it seems that many species would exhibit unrealistic rates of growth for unrealistic periods of time.

For Darwin, the potential problem stemmed from underappreciating a second important thread in his concept of a balance of nature, the vastly complicated and intricately complementary set of ecological interdependencies between species: "how infinitely complex and close-fitting are the mutual relations of all organic beings to each other and to their physical conditions of life" (Darwin 1859, 80). Although most species do not interact directly, Darwin believed they do indirectly through chains of intermediaries. The result is a "web of complex relations" (Darwin 1859, 73) in which species are highly ecologically connected. A specific species' position in the web indicates which other species curb or enhance its growth. Darwin described examples of several such food webs, perhaps the most well known (and engaging) being the ecologically serpentine relation between a clover species (*Trifolium pratense*) and the common cat (Darwin 1859, 73–74). Not all parts of this web and other complex sets of ecological relationships in nature exemplify an antagonistic struggle for survival. Some are beneficial, such as mutualisms benefiting both species. But through those relationships the population-suppressing effects of competitive and predatory struggles are propagated throughout the web.

Unlike previous accounts that assumed a static, providentially predetermined pattern or structure, Darwin's web-based balance of nature concept was rooted in the struggle between individual organisms to survive and reproduce. Species were balanced at their current population levels through a complex array of checks and balances finely honed by natural selection. Darwin emphasized that the exact character of the balance could change as species evolved, so in this sense the so-called niche structure of a community was not fixed. But note that even this kind of balance requires an equilibrium assumption: population levels at a given time reflect the homeostatic processes of a biological community at a point equilibrium. Although this assumption has been supplanted with a recognition that nonequilibrium models with complex dynamics such as chaos, limit cycles, and so-called strange attractors may best represent many types of ecological systems (DeAngelis and Waterhouse 1987), the idea that there is, and perhaps must be, a balance of nature persists. The following sections consider the contemporary account of this balance – characterized intuitively but plausibly by Darwin and early ecologists – with the concept of ecological stability.

## 1 Ecological Stability

With some legitimacy, Arthur (1990, 30) cites the balance of nature as ecology's "number one" research priority, about which there is "near unanimity on its importance" (1990, 35). This priority is recent. Not until the second half of the twentieth century was the concept of a balance of nature rigorously characterized as ecological stability, and predominantly metaphysical speculations about its cause superseded with scientific hypotheses about its basis. But significant uncertainty and controversy remains about which features of an ecological system's dynamics should be considered its stability, and thus no consensus has emerged about how ecological stability should be defined. Instead, ecologists have employed a confusing multitude of different terms to attempt to capture apparent stability properties: "constancy," "persistence," "resilience," "resistance," "robustness," "tolerance," and many more. This, in turn, has resulted in conflicting conclusions about debates concerning the concept based on studies using distinct senses of ecological stability.

One such debate, the stability–diversity debate, has persisted as a (perhaps the) central focus of theoretical ecology for half a century (see Justus 2008a). The debate concerns the deceptively simple question of whether there is a relationship between the diversity of a biological community and its stability. From 1955, when Robert MacArthur (1955) initiated the debate, to the early 1970s, the prevailing view among ecologists was that diversity is an important, if not the principal, cause of community stability. Robert May, a physicist turned mathematical ecologist, confounded this view with analyses of mathematical models of communities that seemed to confirm the opposite, that increased diversity jeopardizes stability. The praise May's work received for its mathematical rigor and the criticisms it received for its seeming biological irrelevance thrust the stability–diversity debate into the ecological limelight, but subsequent analyses have failed to resolve it.

Different analyses seem to support conflicting claims and indicate an underlying lack of conceptual clarity about ecological stability that this section diagnoses and resolves. Below, a comprehensive account of stability is presented that clarifies the concepts ecologists have used that are defensible, their interrelationships, and their potential relationships with other biological properties. In particular, I argue that the concepts of resistance, resilience, and tolerance jointly provide an adequate definition of ecological stability. Roughly speaking, a community exhibits these concepts to a high degree if it does the following: changes little after being perturbed (resistance), returns rapidly to a reference state or dynamic after being perturbed (resilience), and will return to that reference state or dynamic after most perturbations (tolerance).

Besides providing insights about how problematic scientific concepts should be characterized, it is worth noting that the issues involved in characterizing ecological stability have a potential bearing on biodiversity conservation. It seems that for most senses of stability, more stable communities are better able to withstand environmental disturbances, thereby decreasing the risk of species extinction. Positive feedback between diversity and stability would therefore support conservation efforts to preserve biodiversity. This yields a response to an influential criticism. As part of their argument that ecological theory has failed to provide a sound basis for environmental policy, Shrader-Frechette and McCoy (1993) maintain that several proposed definitions of ecological stability are incompatible and that the concept is itself "conceptually confused" or "inconsistent." The account of ecological stability below answers this criticism.

## 2 Benchmarks of Stability: The System Description and Reference State or Dynamic

Stability attributions must be made with respect to two evaluative benchmarks. The first is a system description (*M*) that specifies how the system and its dynamics are represented.[1] The second is a specified reference state or dynamic (*R*) of that system against which stability is assessed. In most ecological modeling, *M* is a mathematical model in which:

1. Variables represent system parts, such as species of a community.
2. Parameters represent factors that influence variables but are (usually) not influenced by them, such as solar radiation input into a community.
3. Model equations describe system dynamics, such as interactions among species and the effect that environmental factors have on them.

[1] In the following, "ecological stability" designates stability of a biological community unless otherwise specified, though most of the discussion also applies to the stability of a biological population or an ecosystem.

$M$ therefore delineates the boundary between what constitutes the system and what is external to it. Relativizing stability evaluations to $M$ is a generalization of Pimm's (1984) relativization of stability to a "variable of interest" because stability is assessed with respect to items (1)–(3) rather than a subset of item (1).

The specification of $M$ partially dictates how $R$ should be characterized, and vice versa. A biological community, for instance, is usually described as a composition of populations of different species. $R$ must therefore reference these populations in some way. For example, $R$ is often characterized in terms of the "normal" population sizes of each species. Since ecological modeling in the late 1960s and 1970s was dominated by the development of mathematically tractable equilibrium models (DeAngelis and Waterhouse 1987), "normal" population sizes were often assumed to be those at equilibrium, that is, constant population sizes the community exhibits unless perturbed. This is not the only possible reference specification, however. A community may be judged stable, for instance, with respect to a reference *dynamic* the populations exhibit. Common examples are a limit cycle – a closed path $C$ that corresponds to a periodic solution of a set of differential equations and toward which other paths asymptotically approach – or a more complicated attractor dynamic (see Kot 2001, chapter 8). Ecological stability can also be assessed with respect to some specified range of tolerated fluctuation. $R$ may also be characterized solely in terms of the presence of certain species.[2] Only extinction would constitute departure from this reference state.

The details of $M$ and $R$ are crucial because different system descriptions – for example, representing systems with different variables or representing their dynamics with different functions – may exhibit different stability properties or exhibit them to varying degrees relative to different specifications of $R$. Specifying $R$ as a particular species composition versus specifying $R$ as an equilibrium, for instance, can yield different stability results. Similarly, different $M$ can produce different assessments of a system's stability properties. Describing a system with difference versus differential equations is one example (May 1974).[3]

[2] To illustrate the partial dependence of $M$ on $R$, notice that the species referred to in $R$ must be part of the system description $M$.

[3] May showed, for instance, that the logistic *difference* equation,

$$N_{t+1} = (1 + r)N_t - \frac{r}{K}N_t^2,$$

where $t$ is time, $r$ is the intrinsic growth rate, $K$ is the carrying capacity, and $N$ is the population size, exhibits dramatically different behavior than the corresponding logistic

Details of $M$ and $R$ are also important because they may specify the spatial and temporal scales at which the system is being analyzed, which can affect stability assessments. Systems with low resistance but high resilience, for example, fluctuate dramatically in response to perturbation but return rapidly to their reference state $R$. Low resistance is detectable at fine-grained temporal scales, but systems may appear highly resistant at coarser scales because their quick return to $R$ prevents detection of fluctuation. Similarly, significant fluctuations in spatially small areas may contribute to relatively constant total population sizes maintained through immigration and emigration in larger regions.

Once (and only once) $M$ and $R$ are specified, the stability properties of a system can be determined. These properties fall into two general categories, depending on whether they refer to how systems respond to perturbation (relative to $R$) or refer to system properties independent of perturbation response. A perturbation of an ecological system is any discrete event that disrupts system structure, changes available resources, or changes the physical environment (Krebs 2001). Typical examples are flood, fire, and drought. Perturbations are represented in mathematical models of communities by externally induced temporary changes to variables that represent populations, to parameters that represent environmental factors, and/or to model structure. Many, perhaps most, real-world perturbations of communities should be represented by changes to both variables and parameters. A severe flood, for instance, eradicates individual organisms and changes several environmental factors affecting populations. In the following, let $P_v$, $P_p$, and $P_{vp}$ designate perturbations that change only variables, change only parameters, and change both, respectively.

Perturbations may cause other changes, such as alteration of the functional form of species interactions, that are not adequately represented by changes to variable or parameter values of typical community models, but which should be included in a comprehensive assessment of community's stability. Since

*differential* equation. For $r > 0$ the logistic differential equation has an asymptotically Lyapunov stable equilibrium $N^* = K$. This is also an asymptotically Lyapunov stable equilibrium of the logistic difference equation, but only for $0 < r < 2$. For $2 < r < 2.526$ the system exhibits a two-period limit cycle. As $r$ increases beyond 2.526 a four-period limit cycle emerges, and the system exhibits chaotic behavior for $r > 2.692$. Thus, although the logistic differential and difference equations appear to describe very similar dynamics, the seemingly inconsequential choice of representing time as a discrete or continuous variable has a substantial effect on evaluating stability properties of the system.

these perturbations change community dynamics, they change $M$. How the altered community responds to these (and subsequent) perturbations must then be assessed against the new description of the community's dynamics as long as those dynamics remain altered. Although a completely adequate assessment of the ecological stability of a community requires consideration of all such changes caused by perturbations, most ecological modeling focuses on changes to variable and parameter values.

## 3 Adequacy Conditions for Stability

There are four plausible adequacy conditions for an account of ecological stability:

(**A1**) The ecological stability of a biological community depends on how it responds to perturbation ([A2]–[A4] specify the form of the required dependency).

(**A2**) Of two communities $A$ and $B$, the more ecologically stable community is the one that would exhibit less change if subject to a given perturbation $P$.

(**A3**) If $A$ and $B$ are in a pre-perturbation reference state or dynamic $R$, the more ecologically stable community is the one that would most rapidly return to $R$ if subject to $P$.

(**A4**) If $A$ and $B$ are in $R$ pre-disturbance, the more ecologically stable community is the one that can withstand stronger perturbations and still return to $R$.

Before considering these conditions in detail, a few remarks help clarify their general basis. First, conditions (A2)–(A4) place only *comparative* constraints on the concept of ecological stability and therefore require only a rank ordering of the stability of biological communities, rather than a particular quantitative valuation. The reason for requiring only comparative constraints is that quantitative valuation of ecological stability depends on the system description ($M$) and reference state or dynamic ($R$) specified for a community, both of which may vary. Second, conditions (A2) and (A3) order the stability of communities based on their behavior following a *particular* perturbation $P$. As adequacy conditions, they therefore do not require a measure of the strength of perturbations. This reflects the difficulties facing the formulation of a measure of perturbation strength (see below), although such a measure is needed to evaluate the resistance of communities when only their responses to perturbations of different strength are known. If a quantitative measure of perturbation strength for different types of perturbation were available, two further *noncomparative* adequacy conditions could be formulated:

(A2′) A highly stable biological community should change little following weak perturbations.
(A3′) A highly stable biological community should rapidly return to its reference state or dynamic following weak perturbations.

In contrast, condition (A4) does require a measure of perturbation strength.

Condition (A1) captures the idea that a community's behavior is a reliable indicator of its ecological stability only if the behavior reflects how perturbation changes the community. If unperturbed, a community may exhibit great constancy throughout some period, for instance, as assessed by a lack of fluctuations in the biomasses of species in the community. It may be, however, that if it had been even weakly perturbed, it would have changed dramatically. Constancy of this community surely does not indicate ecological stability when it would have changed substantially if perturbed slightly. Similarly, variability of a community does not necessarily indicate lack of ecological stability if it is the result of severe perturbations, perturbations that would cause greater fluctuations or even extinctions in less stable communities.

The reason for (A2) is that more stable communities should be less affected by perturbations than less stable ones. Communities that can withstand severe drought with little change, for instance, are intuitively more stable than those modified dramatically. The justification for condition (A3) is that more stable communities should more rapidly return to *R* following perturbations than less stable ones. This adequacy condition captures the idea that lake communities that return to *R* quickly after an incident of thermal pollution, for instance, are more stable than those with slower return rates following similar incidents. The ground for the last condition is that communities that can sustain stronger perturbations than others and still return to *R* should be judged more stable.

## 4 Resistance, Resilience, and Tolerance

Three concepts – resistance, resilience, and tolerance – represent the properties required of ecological stability by conditions (A2′)–(A4). *Resistance* is inversely correlated with the degree a system changes relative to $R$ following a perturbation ($P_v$, $P_p$, or $P_{vp}$). Since perturbations vary in magnitude, resistance must be assessed against perturbation strength. Large changes after weak perturbations indicate low resistance; small changes after strong perturbations indicate high resistance. Resistance is thus inversely proportional to perturbation sensitivity.

Depending on $M$ and $R$, changes in communities can be evaluated in different ways, each of which corresponds to a different measure of resistance. Community resistance is typically measured by changes in species *abundances* following perturbation. It could, however, be measured by changes in species *composition* following perturbation, or in some other way. Pimm's (1979) concept of species deletion stability, for instance, measures resistance by the number of subsequent extinctions in a community after one species is eradicated.

A simple example illustrates the contextual import of $M$ and $R$ in assessing resistance. Consider the classical Lotka–Volterra model of a one-predator, one-prey community:

$$\frac{dx_y(t)}{dt} = ax_y(t) - \alpha x_y(t)x_d(t), \tag{1a}$$

$$\frac{dx_d(t)}{dt} = -bx_d(t) - \beta x_d(t)x_y(t), \tag{1b}$$

where $x_d$ and $x_y$ represent predator and prey populations; $a$ represents prey birth rate; $b$ represents predator death rate; and $\alpha,\ \beta > 0$ in the second term of each equation represent the effect of prey individuals on predator individuals and vice versa. Equations (1a) and (1b) are the description of the system, $M$. There is one nontrivial equilibrium, $x_d^* = b/\beta$ and $x_y^* = a/\alpha$, which is usually specified as the reference state, $R$.

For this $M$ and $R$, resistance to a $P_v$ perturbation that eradicates, say, half of $x_y$ can be measured by how far $x_d$ deviates from $x_d^*$. If $M$ were different, the perturbation could obviously have a different effect on $x_d$. If $x_d^*$ and $x_y$ were competitors, for instance, $x_d$ would increase rather than decrease after this perturbation. Similarly, if $R$ were different, assessments of resistance may change. If $R$ were the species composition $x_d$ and $x_y$ (i.e., $x_d, x_y > 0$) rather than their equilibrium values, for instance, resistance would be assessed in terms of changes from this *composition*, that is, in terms of species extinction. The equilibrium $x_d^* = b/\beta$, $x_y^* = a/\alpha$ is globally stable for this simple community, so only a $P_v$ perturbation strong enough to eradicate one of the species will cause extinction; this model community returns to equilibrium after all other $P_v$ perturbations. For communities with many species and more species dynamics, however, a $P_v$ perturbation that eradicates less than half or half of one species may cause the extinction of that or other species.

Different types of perturbations, moreover, yield different measures of resistance. Since evaluating resistance requires considering perturbation strength, strengths of different types of perturbations must be comparable for there to be a single measure of resistance for a system. Such comparisons are sometimes straightforward. If one perturbation eradicates half of species $x$ in a community, for instance, another that eradicates 75% of $x$ is certainly stronger. If another perturbation eradicates 25% of three species or 5% of fifteen species in the community, however, it is unclear how its strength should be ranked against the perturbation that eradicates 75% of $x$. What criteria could be used to compare strengths of $P_v$, $P_p$, or $P_{vp}$ perturbations, to which systems may show differential sensitivity, is even less clear. Systems that are highly resistant to $P_v$ perturbations may be extremely sensitive to even slight $P_p$. Comparing the resistance of communities is therefore only unproblematic with respect to perturbations of comparable kind.

*Resilience* is the rate at which a system returns to $R$ following perturbation ($P_v$, $P_p$, or $P_{vp}$). Like resistance, resilience must be assessed against perturbation strength unless, although unlikely for many types of perturbation, return rate is independent of perturbation strength. Slow return rates after weak perturbations indicate low resilience, and rapid rates following strong perturbations indicate high resilience. If return rate does not depend on perturbation strength, however, resilience can be evaluated by the return rate independent of the perturbation strength, although the rate may vary across different types of perturbations. Systems may not return to $R$ after perturbation, especially following severe perturbation, so, unlike resistance, resilience is only assessable for perturbations that do not prevent return to $R$. Note that resilience and resistance are independent concepts: systems may be drastically changed by weak perturbations (low resistance) but rapidly return to $R$ (high resilience), and vice versa.

Resilience is commonly measured as the inverse of the time taken for the effects of perturbation to decay relative to $R$. For a specific mathematical model, this can be determined analytically or by simulation. For the community described by Equations (1a) and (1b) above, for instance, resilience to a $P_v$ perturbation that eradicates half of one species could be simply measured by $\frac{1}{|t_{eq} - t_p|}$ where $t_p$ is the time at which the community is initially perturbed and $t_{eq}$ is the time at which the community reestablishes equilibrium. Resilience to $P_v$ perturbation is determined by the largest real eigenvalue part for systems modeled by linear differential equations if it is negative, and analytic methods have been developed to assess resilience to $P_v$ perturbation for nonlinear models. Empirical measurement of resilience for communities in nature, however, is often thwarted by subsequent perturbations that disrupt return to $R$. This difficulty can be avoided if subsequent perturbations can be evaded with controlled experiments. If the return rate is independent of perturbation strength, estimation of resilience is also more feasible because only the decay rate of the perturbation effects need be measured before the system is further perturbed; measurement of perturbation strength is not required (Pimm 1984, 1991). Like resistance, furthermore, different types of perturbations yield different measures of resilience since return rate to $R$ may depend on the way in which systems are perturbed. A system may be highly resilient to $P_v$ perturbation and poorly resistant to $P_p$ perturbation, for instance, or more resilient to some $P_v$ or $P_p$ perturbations than others.

*Tolerance*, or "domain of attraction" stability, is the ability of a system to be perturbed and return to $R$, regardless of how much it may change and how long its return takes. More precisely, tolerance is positively correlated with the range and strength of perturbations a system can sustain and still return to $R$. The magnitudes of the strongest perturbations it can sustain determine the contours of this range. Note that tolerance is conceptually independent of resistance and resilience: a system may be severely perturbed and still return to $R$ (high tolerance), even if it changes considerably (low resistance) and its return rate is slow (low resilience), and vice versa.

Similar to resistance and resilience, different kinds of perturbations yield different measures of tolerance. Tolerance to $P_v$ perturbations, for instance, is determined by the maximal changes variables can bear and not jeopardize the system's return to $R$. With respect to $P_v$ perturbations that affect only one species of a community, for instance, tolerance can be simply measured by the proportion of that species that can be eradicated without precluding the community's return to $R$. If a nontrivial equilibrium of Equations (1a) and (1b) from above is globally stable, for instance, the community described by the equation is maximally tolerant to $P_v$ perturbations relative to this reference state because the community will return to it after any $P_v$ perturbation that does not eradicate one of the species. Variables of a system may be perturbed, however, in other ways. A $P_v$ perturbation may change all variables, several, or only one; may change them to the same degree; may change some variables more severely than others; and so on. How exactly variables are perturbed may affect whether the system returns to $R$. System tolerance must therefore be evaluated with respect to different types of perturbation. The same goes for assessing tolerance to $P_p$ or $P_{vp}$ perturbations.

## 5 Explicating the Balance of Nature

Although resistance, resilience, and tolerance do not adequately explicate ecological stability individually, they do so collectively. In fact, they constitute jointly sufficient and separately necessary conditions for ecological stability, notwithstanding Shrader-Frechette and McCoy's (1993, 58) claim that such conditions do not exist. Consider sufficiency first. Since these three concepts represent the properties underlying conditions (A2)–(A4) (and [A2′] and [A3′]), communities exhibiting them to a high degree would change little after strong perturbations ([A2]), return to $R$ rapidly if perturbed from it ([A3]), and return to $R$ following almost any perturbation ([A4]). If $R$ is a point equilibrium, moreover, a community exhibiting high resistance, resilience, and tolerance will be relatively constant. As such, these three properties certainly capture ecologists' early conceptions of ecological stability, and there seems to be no further requirement of ecological stability that a community exhibiting these properties would lack.

Each concept is also necessary. Highly tolerant and resistant communities, for instance, change little and return to $R$ after most perturbations. In regularly perturbing environments, however, even a highly resistant and tolerant community may be iteratively perturbed to the boundary of its tolerance range and "linger" there if its return rate to $R$ is too slow. Subsequent perturbations may then displace it from this range, thereby precluding return to $R$. If this community rapidly returned to $R$ after most perturbations (high resilience), it would rarely reach and would not linger at its tolerance boundary. In general, low resilience preserves the effects perturbations have on communities for extended, perhaps indefinite durations, which seems incompatible with ecological stability.

Similar considerations show that tolerance and resistance are necessary for ecological stability. A highly resilient and tolerant but weakly resistant community rapidly returns to *R* following almost any perturbation, but changes significantly after even the slightest perturbation, which seems contrary to ecological stability. The dramatic fluctuation such communities would exhibit in negligibly variable environments is the basis for according them low ecological stability. A highly resilient and resistant but weakly tolerant community changes little and rapidly returns to *R* when perturbed within its tolerance range, but even weak perturbations displace it from this range and thereby preclude its return to *R*, which also seems contrary to ecological stability.

Resistance, resilience, and tolerance are independent concepts, and thus biological communities may exhibit them to different degrees. Although the necessity of each concept for ecological stability does not strictly entail that they are equally important in evaluations of a community's stability, nothing about the pre-theoretic concept of ecological stability seems to suggest otherwise. As a concept composed of resistance, resilience, and tolerance, ecological stability therefore imposes only a partial, not complete, ordering on communities. Moreover, since communities may differentially exhibit resistance, resilience, and/or tolerance for different types of ecological perturbations, each property also imposes only a partial ordering on communities. This twofold partiality entails that inferences from analyses of stability–diversity and stability–complexity relationships are limited by the property and type of perturbations analyzed, beside the particular system description (*M*) and reference state or dynamic (*R*) specified.

It is worth pausing over what the framework for ecological stability presented above shows about the general concept. It certainly shows that ecologists have used the term "stability" to describe several distinct features of community dynamics, although only resistance, resilience, and tolerance adequately define ecological stability. This plurality does not manifest, however, an underlying vagueness, "conceptual incoherence," or "inconsistency" of the concept, as Shrader-Frechette and McCoy (1993, 57) suggest in their general critique of basic ecological concepts and ecological theories based on them.

Two claims seem to ground their criticism. First, if "stability" is used to designate distinct properties, this indicates the concept is itself conceptually vague and thereby flawed. Although terminological ambiguity is certainly undesirable, most ecologists unambiguously used the term to refer to a specific property of a community and accompanied the term with a precise mathematical or empirical operationalization. Since these were in no sense vague, in no sense was ecological stability "vaguely defined" (Shrader-Frechette and McCoy 1993, 40). Ecologists quickly appreciated this terminological ambiguity, moreover, and began explicitly distinguishing different senses of ecological stability with different terms (Odenbaugh 2001).

Shrader-Frechette and McCoy's second claim is that "There is no homogeneous class of processes or relationships that exhibit stability" (1993, 58). The underlying assumption seems to be that concepts in general, ecological stability in particular, must refer to a homogeneous class in order to be conceptually unproblematic. That ecological stability does not, and worse, that ecologists have supposedly attributed inconsistent meanings to it, shows the concept is incoherent, they believe, much like the vexed species concept (1993, 57). Shrader-Frechette and McCoy do not offer an argument for this assumption, and it is indefensible as a general claim about what concepts must refer to. Common concepts provide clear counterexamples. The concepts "sibling," "crystal," and "field," for instance, refer to heterogeneous classes, but there is nothing conceptually problematic about them. There is debate about the idea of disjunctive *properties* in work on multiple realization (Kim 1998), but criticisms raised against disjunctive properties do not necessarily apply to disjunctive *concepts*, nor were they intended to. Kim (1998, 110) emphasizes this point:

> Qua property, dormativity is heterogeneous and disjunctive, and it lacks the kind of causal homogeneity and projectability that we demand from kinds and properties useful in formulating laws and explanations. But [the concept of] dormativity may well serve important conceptual and epistemic needs, by grouping properties that share features of interest to us in a given context of inquiry.

Even if criticisms of disjunctive properties were sound, it therefore would not follow that the disjunctive concepts such as ecological stability are also problematic. Shrader-Frechette and McCoy's criticism of ecological stability is therefore indefensible. The conceptual and epistemic utility of a concept is enhanced, furthermore, if clear guidelines for its application exist, which the above analysis has attempted to provide.

Moreover, the definitional status of the concepts of ecological stability and species is not analogous. Biologists have proposed plausible, but incompatible competing definitions for the species concept because it is problematically ambiguous (Ereshefsky 2001; Reydon 2013). That resistance, resilience, and tolerance have been referred to under the rubric "stability," however, does not show that ecological stability is similarly problematically ambiguous because they are conceptually independent and therefore compatible, as different senses of "species" arguably are not.[4] In addition, most ecologists recognized that there are several components of ecological stability, and individual stability concepts such as resistance, resilience and tolerance, or measures thereof, were rarely proposed as *the* uniquely correct definition of ecological stability. Rather, they were and should be understood as distinct features of ecological stability or ways of measuring it, not competing definitional candidates. Like many scientific concepts, ecological stability is multifaceted, and the distinct referents ecologists attributed to it accurately reflect this. Conceptual multifacetedness alone does not entail conceptual incoherence or inconsistency.

[4] As noted above, different quantitative measures of resistance, resilience, and tolerance may be incompatible. This does not establish, however, that the corresponding concepts are incompatible.

# 4 Modeling in Ecology

## Representation and Reduction

To study the biological world, ecologists must do two things. One, they must conduct the fieldwork – experimental manipulation, monitoring, sampling, surveys, and so on – that furnishes the raw observational information at the base of all scientific knowledge. Ecology, like any other empirical science, is an a posteriori enterprise; without basic observational data nothing (epistemically defensible) can be said. But the insights that can be gleaned from this data alone are limited. The second thing ecologists must do, to harness data's full epistemic power, is to create a model, a representation of the portion of the biological world of interest. Only with such a representation can observational information be integrated and systematized into a more comprehensive and detailed account of how the biological world works.[1] *Perhaps* it is not as epistemically foundational, but modeling is just as indispensable as observation to achieving significant ecological insights and ultimately building successful general theories in ecology.

By far the most common type of representation in ecology, and so-called hard sciences generally, is mathematical modeling. There are of course other kinds of models. Visual, physical, diagrammatic, and other types of models can play important roles in ecology, and they certainly figure prominently in other sciences. Visual and diagrammatic modeling in cognitive science (Cummins 1996) and concrete physical modeling in chemistry and engineering (Schaffner 1969a, 1969b) and "model organisms"

[1] That is not to deny that observational information by itself cannot facilitate important scientific insights. It clearly can. But theories are simply models by another name (and perhaps with a more ambitious scope), and the power of either to unlock the true potential of observational data is unquestionable (see Godfrey-Smith 2006; Weisberg 2013).

in biology (Ankeny 2001; Ankeny and Leonelli 2011) are paradigmatic examples. As in many sciences (Chandrasekharan and Nersessian 2015), diagrammatic reasoning is also often a crucial step in the development of more precise mathematical models in ecology, particularly ecosystem ecology. And there are methods of qualitative diagrammatic modeling, such as loop analysis (see Puccia and Levins 1985), that are just as rigorous as fully quantitative methods of mathematical modeling (Justus 2005) and that can generate significant insights about the ecological systems they target (Elliott-Graves forthcoming). The imprecise nature of these modeling methods can even be an asset. It can enhance generality by helping ensure the results they uncover hold across a wide variety of qualitatively similar but quantitatively different systems. But imprecision can also obviously be a liability, imposing limitations on what insights can be acquired about the systems the models are intended to represent (Taylor and Blum 1991; Justus 2006). Correctly or myopically, fully quantitative mathematical modeling has received the lion's share of attention and theoretical aspiration from ecologists.

Mathematical models in ecology are employed to represent and understand populations, communities, ecosystems, metapopulations, metacommunities, and many other biological systems. The most common, and analytically well developed, are models of populations and communities. Given the vastly different scales and diversity of representational targets, it is unsurprising these models exhibit a wide array of functional forms and contain many distinct types of variables and parameters. This complexity and the sometimes uneasy relationship between theoretical and empirical work in ecology (see Worster 1994; Kingsland 1995) connects ecology to the extensive literature in philosophy of science on modeling and scientific representation.[2] Rather than simply retrace the well-worn themes of that literature with ecological examples, this chapter focuses on two subjects with uniquely ecological content and significance.

One critical subject is how the theoretical assumptions, analogies, and idealizations that feature prominently in ecological modeling should be conceptualized and justified. Most ecological models are *highly* idealized. That is, they incorporate unrealistic assumptions – for example, they ignore some system components and interactions, treat interactions as

[2] See Weisberg 2013.

*instantaneous* and assume their effects propagate similarly, represent discrete components with continuous variables, or describe community structure nonspatially – to make model analysis tractable. Unrealistic idealizations, however, make it uncertain whether modeling results demonstrate properties of the represented system or are byproducts of the idealizations. Since it is often unclear which properties are primarily responsible for system dynamics, idealizations may significantly mischaracterize its most important features. Consequently, an enhanced sense of understanding conveyed by an idealized model may fail to be about the system it is intended to represent.

This misalignment can be especially troubling in ecology because ecological modelers often uncritically emulate mathematically sophisticated models of physics to ensure their modeling is rigorous. The tacit assumption motivating the emulation is often the belief that the dynamics of ecological systems are sufficiently analogous to those studied in physics. Using a well-known concept from mathematical modeling in physics, Lyapunov stability, to analyze the putative stability of ecological systems is one example. Another is the fascinating idea, explicitly and cogently defended rather than implicitly assumed, that there is a close analogy between the second-order dynamics of Newtonian mechanics and the dynamics of biological populations originally proposed by the ecologist Lev Ginzburg (1972, 1986) and later elaborated with the philosopher Mark Colyvan (Ginzburg and Colyvan 2004).

Another critical subject is reduction in ecology. Populations are the basic units of classical models of theoretical ecology. In them, familiar ecological interactions are represented as population-level processes. How individual organisms actually interact is typically ignored in favor of state variables designating population-level properties such as intrinsic growth rate, carrying capacity, or (average) competition and predation pressure. *Individual-based models* (IBMs) are different. They do not aggregate over, or abstract from, the details of interactions between individual organisms. For example, a typical IBM is a complex computer-based mathematical model composed of a set of variables for each organism in the system being modeled and sets of equations representing how each organism consumes resources, develops, interacts with its environment and neighbors, reproduces, and, when applicable, rears its offspring. IBMs are incredibly complicated, often involving thousands of variables, parameters, and equations of a variety of mathematical forms.

Their dynamics are therefore not analytically tractable and must be investigated by simulation techniques.

That IBMs are not analytically tractable is considered a serious deficiency. But their greater realism – that they explicitly incorporate the individual-level processes that plausibly drive higher-level ecological dynamics – constitutes a significant epistemic advantage. Ecological systems have intricate internal structure that only IBMs seem to accurately capture and that classical population-level models seem to obscure. Individual organisms do not all possess the same phenotypic characteristics or play the same ecological roles, nor obviously do they share equivalent life histories. Rather, interactions between them are spatially and temporally local and depend significantly on precisely these idiosyncrasies.

Individual-based modeling in ecology is but part of a more general, well-vetted individualistic approach to modeling the world frequently called *methodological individualism* (MI) that has been utilized extensively and fruitfully in the social sciences (Heath 2015). This chapter assesses the ecological promise of IBMs and the philosophical issues they pose. In particular, two central features of IBMs yield significant scientific merits: representational flexibility and the reduction of the population level to the individual level. The former concerns the greater expressive power of IBMs over classical models. With IBMs, for example, latitude exists to incorporate crucial ecological details such as spatial and age-class structure, differential mobility and resource utilization, and phenotypic variability. Effective testing of ecological theory seems to require this flexibility. IBMs also seem to demonstrate how properties of populations and communities emerge from processes in which individual organisms uniquely participate; the former therefore reduces to the latter. Although enticing, these claims are controversial and require close scrutiny, toward which we now turn.

## 1 Reduction in Ecology

Cross-pollination between theorizing in biological and social sciences has a long and fruitful history, most of which explores connections between rational choice and evolutionary theory. The usual focus is how evolutionary concepts and principles illuminate models of the economic world, and vice versa. Evolutionary economics exemplifies the former; evolutionary game theory and optimal foraging theory, the latter. Recent philosophical work targets the same confluences: evolutionary origins of the social contract (Skyrms 1996), risk aversion, and analogies between fitness and utility (Okasha 2011), among others. The veritably exclusive focus on evolution, however, is grievously partial. Ecology is at the heart of our understanding of the biological world, and an ecological perspective is as indispensable to evolutionary theorizing as evolution is to ecology. Fruitful cross-fertilizations of social-scientific and ecological theorizing should therefore be expected, but surprisingly remain quite underexplored.

The MI manifested in individual-based modeling in ecology is part of that untold story. In the social sciences MI constitutes a constraint on how social phenomena should be conceptualized and represented. It means different things to different thinkers, but privileging individuals in explanations, representations, and theories of higher level phenomena is a common denominator (Heath 2015). In ecology, MI finds clear expression in a relatively new modeling strategy: individual-based modeling (see Grimm and Railsback 2005). Just as expected utility calculations and actions of individual agents comprise the preferred level of analysis according to MI in social sciences, behaviors and physiologies of individual organisms function similarly for ecological individual-based models (IBMs). This section first describes and defends this analogy, and then evaluates how much light it sheds. Debates about MI in the social sciences reveal important insights about how individual-based ecological modeling should be understood. For example, IBMs need not be construed as ontologically reducing population or community-level properties to individual-level properties, or committed to any metaphysical view about what marks the "real."[3] The explanatory and representational priority on individuals, and the compelling rationale underlying it, buttresses the prospects of IBM-based reduction in ecology.

The next section describes significant differences between classical and individual-based ecological modeling. Section 1.2 considers confusions about methodological individualism and the particular way it aligns with reductionism. Sections 1.3 and 1.4 clarify how a representational priority on individuals yields a cogent albeit deflationary account of ecological emergence and can underpin the unification of disparate models and theories in ecology.

[3] For the metaphysically inclined, however, this baggage can of course be added.

### 1.1 *Classical versus Individual-Based Modeling in Ecology*

Almost every introduction to ecological modeling begins with the canonical Lotka–Volterra one-predator, one-prey, community model:

$$\frac{dV}{dt} = [\text{prey growth rate}] - [\text{capture rate of prey per predator}]P, \tag{1a}$$

$$\frac{dP}{dt} = [\text{predator births per capture}]P - [\text{predator death rate}], \tag{1b}$$

where $V$ and $P$ designate prey and predator abundances. Besides the pedagogical virtues of starting simple before tackling more sophisticated representations, Equations (1a) and (1b) exemplify standard modeling strategy in ecology. Basic units are biological populations, and ecological processes such as predation and competition are typically represented with differential or difference equations *at the population level*. When available, information about properties of individual organisms is statistically aggregated into population-level state variables representing birth, death, capture rates, and other factors. More often than not it is unavailable, in which case state variables simply abstract away from individual-level details.

The approach has several advantages. It situates ecological modeling in a mathematically precise, rigorous, and well-developed analytic framework that has proved fruitful in many sciences: dynamic systems theory. Theorems and derivational techniques within this framework constitute powerful tools for analyzing systems in other domains – classical and fluid mechanics, chemical kinetics, and many more – and are thought to do the same for ecology (see Justus 2008a). Second, abstracting from the individual level not only accommodates the relative paucity of available ecological data on individual organisms and their interrelations; it better facilitates general theorizing about population and community dynamics. Rather than become mired in sundry details about individuals, focusing on state variables seems to offer the best prospects for uncovering generalizations that hold across different configurations of individuals into populations and communities. Without the abstraction, it seems unlikely highly regarded fruits of ecological theorizing such as the competitive exclusion principle or island biogeography theory would have been found.

Classical ecological models are typically highly idealized and unrealistic. For example, prey only die by predators consuming them and predators only increase by converting prey into births in Equations (1a) and (1b).

More sophisticated models better avoid such conspicuously unrealistic assumptions, but ignoring individuals means sacrificing maximal realism. This need not detract from the insights the models provide. Besides the tractability simplifying assumptions afford, they also help make representations of complex systems cognitively evaluable, especially in ecology (Odenbaugh 2005). Aggregating or abstracting from individual-level details also seems reasonable since organisms of a single species share most properties in virtue of that fact.

It is increasingly being appreciated, however, that classical models overlook two significant drivers of ecosystem dynamics: variability between organisms, even of the same age and species, and locality of interactions, both spatially and temporally. Classical models are blind to these features; IBMs are not. This significant advantage was recognized early by supporters of IBMs: "The assumption that all individuals are equally affected by increasing population density disagrees with empirical evidence ... and leads theoretical population biology into a blind alley" (Łomnicki 1978, 473).

Organisms rather than populations are the basic units of IBMs, and organismal behavior, development, reproduction, and interactions are represented *explicitly at the level at which they occur*. A *set* of variables represents each individual, and a *set* of equations represents relations between them and abiotic factors such as precipitation, temperature, or soil acidity. These variables and equations are intended to reflect the detailed physiologies and spatially specific interactions usually absent from classical models and, crucially, their variance across individuals. Biologically, organisms of a species are not created equal, and IBMs capture this important fact. And with spatial and temporal dynamics explicitly represented, the effects, for instance, of neighborhood relations and the locality of environmental gradients on overall ecosystem dynamics – individuals of those species are adjacent to this individual, that region is sunny and dry while this one is shaded and moist, and so on – can thereby be investigated. Admittedly, this attention to detail has a cost. IBMs contain hundreds, sometimes thousands, of variables and equations linking them, and are therefore analytically intractable, solvable only by simulation.

Despite this computational hurdle, some ecologists believe that integrating the individual-level information lost with state variables lends a crucial advantage to individual-based modeling. A recent comprehensive introduction to IBMs, for example, claims that "Individual-based models have demonstrated the potential significance of individual characteristics to population dynamics and ecosystem processes. Even in its infancy, individual-based modeling (IBM) has changed our understanding of ecological systems" (Grimm and Railsback 2005, xiii). Others are much less impressed (see Roughgarden 2012).

To adjudicate this emerging controversy, a much clearer understanding is needed of the modeling strategy IBMs employ, its epistemic credentials, and potential deficiencies. Fortunately, IBMs instantiate a general individualistic approach to representing the world previously mentioned as "methodological individualism" (MI) that has been applied extensively and fruitfully in other sciences, particularly social sciences, and over which much ink has been spilt. Extant debates about its legitimacy offer insights about how individual-based ecological modeling should be conceptualized. But as Kincaid (1997, 13) aptly observes, debates about MI have been "long on rhetoric and short on clarity." The next section navigates through potential confusions to pinpoint the defensible sense in which IBMs embody MI.

## 1.2 *Methodological Atomism, Individualisms, and Reductionism*

MI is but one of numerous individualisms. Defensibly construed, it does not require *ontological individualism*, the idea that only individuals exist and aggregates, collections, and other wholes do not, or do not somehow exist "over and above" their individual constituents. Quine's forceful arguments notwithstanding, methodological commitments within scientific modeling, or epistemic inquiry generally, do not necessitate a particular thread in the tangle of metaphysical issues concerning ontology (see Azzouni 2004).

Nor is MI *methodological atomism*, the idea that wholes are explainable only by properties of individuals, not their interrelations. This view – sometimes attributed to Hobbes's account of how the civil state emerges from "solitary," pre-social creatures in the state of nature (Heath 2015) – would disallow even binary, pairwise interactions invoked in microeconomics to explain macroeconomic patterns (Hoover 2010). Interactions between individuals and their environments that drive ecosystem dynamics are similarly at the core of ecological IBMs.

This inclusivity prompts a clarification and raises important issues. First the clarification. Within social sciences, MI prioritizes individual properties and actions, interactions they catalyze, and ultimately the underlying intentional states motivating both. These supply the base from which all social phenomena can be accounted for according to MI. Methodological atomism is more restrictive: the origin and content of these intentions must be individualistically pure. They might arise from individual-level selective history, developmental processes, percolating quantum indeterminacy, or something entirely sui generis, but broader sociological, political, cultural, or other nonindividualistic factors are prohibited. For example, aspirations of increasing the general welfare of others or fear of societal decay, both of which reference extra-individual social factors, cannot be responsible for individuals' intentions and subsequent actions for methodological atomists. If they were, higher-level factors would contaminate the unadulterated lower-level individualistic basis that supposedly accounts for social phenomena. Similarly, interactions cannot be invoked to account for social phenomena, unless perhaps they follow trivially from the properties of individuals atomistically conceived. Invoking associative or neighborhood effects caused (even partially) by broader social dynamics is therefore disallowed.

Despite assertions from some of its critics,[4] plausible versions of MI are unencumbered by this utterly implausible view of human psychology. Besides representing individuals explicitly, MI endeavors to represent them realistically. Social, political, and cultural factors obviously influence individuals' beliefs, intentions, and actions, and MI can correctly accommodate that fact.[5] The lesson is the same in individual-based ecological modeling. Properties organisms possess reflect selection histories that depend on particular past environments and fitness landscapes, both of which population dynamics influence. Physiologically onerous male traits of many species – the marvelous spatuletail, peacock, Irish elk – display the selective power of such dynamics. And apart from evolutionary considerations, ecological dynamics exhibit similar dependencies. Organisms' properties, physiologically and behaviorally, can differ dramatically based on status and pedigree in

[4] Sen (2009, 244), for example, endorses the characterization of MI as "all social phenomena must be accounted for in terms of what individuals think choose and do," and then says, "There have certainly been schools of thought based on individual thought, choice and action, *detached from the society in which they exist*" (emphasis added). Only the former defensibly characterizes MI, and it differs from and does not entail the latter.

[5] Professed advocates of MI too frequently ignore these influences by uncritically endorsing unrealistic accounts of human motivation and rationality. I follow Heath (2015) in sharply distinguishing MI from such views.

populations, particularly as cognitive sophistication and social complexity increase. In a congress of orangutans, for example, alpha and beta males of similar age and heredity look and act very differently.

Beyond the origin of individual properties and intentional content, realistically representing even binary interactions between individuals also requires rejecting the stringency of methodological atomism because these interactions are often shaped by extra-individualistic factors. For example, broad structural features of populations can partially determine which conspecifics organisms associate with and how, something that would not follow from their dubiously conjectured atomistic properties. But this prompts a difficult question: if representing binary interactions with these influences is permissible for MI, are higher-order $n$-ary relations as well? Realism seems to again dictate yes, at least for some $n > 2$. In humans and simians, for example, instances of complex, higher-order relationships abound. That less cognitively endowed organisms sometimes exhibit similar complexity is less appreciated. Fire ants (*Solenopsis invicta*) in North America typically outcompete native ant species to local extinction. But if even low densities of parasitoid flies from their native range are present, *Solenopsis* consumes less, grows to smaller size, and competes less effectively against native ants (Mehdiabadi and Gilbert 2002). Accurately representing these interactions seems to require ternary relationships, and it is an open empirical question whether and how often higher-order relations are necessary to represent systems across different scientific domains. This presents a potential impediment: although it also seems entirely arbitrary to posit that an $n$ exists above which MI is abandoned, as $n$ increases, recognizing $n$-ary relations surely departs from an individual-level focus. Whether this recognition would necessitate rejecting MI depends on the ultimate origin of those relations (see below).

Such permissiveness raises another concern: it ostensibly precludes reducing higher-level phenomena. "Methodological individualism" and "reductionism" are frequently considered synonymous labels (Kincaid 1997). At the very least, reductionism is taken to require MI. But if MI condones appropriating extra-individualisitic information when modeling systems, this seems to preclude rather than facilitate reduction.

An unnecessarily strong conception of MI underlies these concerns. Unlike methodological atomism, the epistemic priority MI places on individuals does not necessitate that particular representations of systems be individualistically pure, free from *any* influence of extra-individualistic factors. Such a stringent criterion is tantamount to stipulating that mathematical proofs contain logical derivations for every proposition they utilize, however foundational or well established. For proofs in all but a small subset of issues in the most foundational fields, the goal is not to establish theorems from the axiomatic ground up. Rather, specific proofs establish new theorems conditional on others, some of which may be controversial or even actively doubted. IBMs function similarly for reductionist programs. They establish how properties and interactions between individuals, which in turn may depend on input from extra-individualistic sources, can account for higher-level phenomena. Whether the phenomena can be *fully* accounted for ultimately depends on whether this input can itself be accounted for individualistically. In cases with such input, IBMs therefore facilitate a kind of *conditional reduction*: conditional on the relevant extra-individualistic features being reducible, IBMs show how reductionism's front can advance. A genuine exception to MI and reductionism obtains only if they cannot.

Whether these features are ultimately reducible is an exceedingly difficult context-dependent and, in part, empirical question that the current state of scientific knowledge often gives only dim glimpses of. Occasionally, it is relatively clear. An institution's culture may shape the beliefs, intentions, and actions of agents within it, and influence how and with whom they interact. But tracing an institution's origin back to a charter among individual founders and the subsequent history of individual decisions responsible for its culture presumably eliminates this higher-level influence. There is likely a reminder, of course, in the legal and governmental framework without which charters could not exist, but no insurmountable obstacle appears to preclude iteratively adopting precisely the same approach, at least in principle. Practical priorities may make such monumental undertakings an unwise allocation of limited resources, but pragmatic considerations do not challenge the plausibility of these social entities and phenomena ultimately being accounted for individualistically. Individualistic accounts of how monetary institutions emerge from bartering interactions as much more efficient exchange mechanisms are a widely heralded MI success (see Nozick 1977).

Social phenomena that are not purposive artifacts of human planning, such as widely shared fairness and equality norms, pose a more formidable challenge. They likely pre-dated legal and political institutions, and thereby elude the MI strategy just described. They do not elude, however, a wide range of "bottom-up" modeling techniques based on replicator dynamics.[6] Replicator dynamics constitutes a simple model of evolution in which populations change through differential reproduction or differential imitation for cultural evolution. The approach does not necessitate individuals be represented explicitly, but, intriguingly, when individual-level dynamics are included modeling results often become more plausible. Alexander and Skyrms (1999), for example, analyze potential evolutionary origins of the widespread sense of distributive justice with two-player divide-the-dollar bargaining games. Each player formulates a bottom-line demand about how a dollar should be distributed. If demands jointly exceed $1, no one gets anything; otherwise, each gets what was demanded. Game-theoretic experiments overwhelmingly result in fair, 50-50 splits, which classical rational choice theory cannot explain. Standard replicator dynamics, which assume individuals randomly pair to bargain from a very large population, offers a better explanation. Fair division goes to fixation more often than not, and the farther other divisions are from fair the lower their probability. But by far the best results emerge when locality is imposed on interactions. When individuals bargain only with their eight-member Moore neighborhood rather than randomly, fair division almost always goes to fixation, which closely fits empirical facts, and it fixes much more rapidly. This provides a striking glimpse of how social and moral norms might emerge from individual-level dynamics. Much more remains to be explored with these techniques, but they have already uncovered tantalizing suggestions about how lower-level dynamics, even among rudimentary creatures, might generate primitive social norms (Skyrms 1996), representational information (Skyrms 2010), category formation (Skryms 2010), and others. The next section explains how emergence occurs similarly in ecology. Of course, endorsing MI does not require committing to this modeling strategy's universal applicability or success. But its successes to date demonstrate that the dependency of some individual-level properties and interactions on higher-level factors poses interesting questions for future research, rather than reveals insurmountable obstacles to reduction.

A final clarification: MI's ambition is often overstated. MI maintains that non-individualistic explanations are often inferior, incomplete, and

[6] See Skyrms 1996 for an accessible introduction.

therefore too often inadequate, but they are not thereby specious or necessarily useless. Regarding explanation, for instance, MI need not commit to the misguided view that wholes are *only* explainable by properties of individuals and their interactions with each other and the environment. This is only attractive on an implausible view that there is a uniquely correct explanation, or kind of explanation, for a given explanandum. The same holds for model representation. Explanations and models serve a variety of ends, and some well-established ends – enhancing understanding, empirical adequacy, discovering underlying mechanisms, or expanding a theory's scope – are served in a variety of ways. "Science has room for both lumpers and splitters," as Sober (1999, 551) deftly put it.

Having emerged from a conceptual thicket of potential confusion with a clearer understanding of MI, its plausibility and merits in ecology can be assessed. Section 1.3 proceeds from the most plausible and least controversial points of support to more contentious issues of emergence.

## 1.3 Supervenience, Compositionality, and Emergence in Ecology

*Supervenience.* Although methodological doctrines do not entail metaphysical ones, the latter can bear on the former. If one domain fails to supervene on another, for example, the latter presumably cannot be an adequate methodologically individualistic basis for the former. As predicated about relations between domains or sets of properties or facts, and characterized most plainly, A supervenes on B if and only if any difference in A entails a difference in B: no A-difference without a B-difference (see McLaughlin and Bennett 2018 for a comprehensive review). Since one domain's facts would not determine those of another, some facts of the nonsupervening domain would be unascertainable via the other. MI has been criticized in the social sciences on this ground (Epstein 2012), but no alleged obstacle there threatens the claim that population and community facts supervene on properties of individual organisms, their interrelations, and relations with the environment.

*Compositionality.* Undergirding supervenience in ecology is another uncontroversial claim: individual organisms constitute biological populations and communities.[7] This compositionality removes a gap frequently

[7] This uncontroversial claim is largely unrelated to contentious issues about how population concepts should be characterized (see Millstein 2009). The latter primarily

thought to pose insuperable obstacles to reduction in other contexts: the inability to bridge fundamentally different domains. As lodged against reductive aspirations, such gaps have been maintained for connections between mind and body, psychology and neurobiology, first-person and third-person perspectives, classical and molecular genetics, and others (see van Riel and Van Gulick 2019 for a comprehensive review). And even when these domains are taken to be systematically related, the nature of the relationship is usually conceptualized as instantiation, realization, or simple correlation, rather than mereologically. Ecology's purview fortunately does not span fundamentally different domains, so it happily avoids an intractable "hard problem."[8]

*Emergence.* Supervenience and compositionality are uncontroversial in ecology. Emergence is not. Though far from consensus, some population, community, and ecosystem-level properties are considered emergent and thus impervious to reductive analysis (see Drake et al. 2007 for a sampling). A survey is impossible here, but one prominent, well-studied candidate for ecological emergence that has garnered philosophical attention illustrates underappreciated merits of MI outside the social sciences, and reveals particularly clearly how IBMs can contribute to reductive (and deflationary) accounts of putatively "emergent" properties in ecology: community stability.

As Chapter 3 recounted, the concept has a long history, tracing back at least to Aristotle's ideas about a balance of nature, if not earlier (Egerton 1973). It was formulated in precise terms, however, only in the mid-twentieth century with seminal work by Robert MacArthur and Robert May, among others (see Justus 2008b). Their characterizations differed, and that pluralism continues today. But among the diversity of notions, one definition initially developed by David Tilman (1999) to assess the stability

concern what kind and how strong relations must be between conspecifics (causally, genealogically, or spatiotemporally) for a defensible population concept to apply, rather than whether populations are composed of conspecifics. Indeed, all definitions of population concepts in that debate affirm compositionality. Note that an analogous compositionality claim is also plausible for ecosystems: they are composed of individual organisms and their abiotic environments.

[8] One might think this judgment is too quick. Perhaps compositionality can fail when domains are not considered fundamentally different. Kincaid (2004, 302) suggests some social entities, such as "society," might contain non-individuals, such as "capital." The institutions of "money" and "property" might also be candidates. Irrespective of whether these should be considered proper societal parts or simply byproducts of interactions between parts, biological populations and communities seem to contain no counterparts that challenge compositionality.

of grassland communities is widely entrenched, particularly among field ecologists. His concept, labeled "temporal stability," focuses on how population biomasses in communities vary over time. In particular, temporal stability decreases as biomass variability increases, the motivating idea being that if two series of abundances are plotted across time, the more stable one exhibits less fluctuation. Measuring temporal stability therefore requires empirical measurement of biomasses over some period. More precisely, let $B_i$ be a random variable designating the biomass of species $i$ in an $n$-species community $C$ assayed during some time period, and let $\overline{B_i}$ designate the expected value of $B_i$ for that period. Temporal stability $S_t$ of $C$ is then:

$$S_t(C) = \frac{[mean\ species\ biomass]}{\sqrt{[biomass\ variance] + [biomass\ covariance]}}$$

$$= \frac{\sum_{i=1}^{n} \overline{B_i}}{\sqrt{\sum_{i=1}^{n} Var(B_i) + \sum_{\substack{i=1;j=1 \\ i \neq j}}^{n} Cov(B_i, B_j)}}, \tag{2}$$

where *Var* and *Cov* designate variance and covariance. The denominator measures how much the populations comprising a community fluctuate over some period; the numerator measures the expected mean biomasses of those populations. With this equation, $S_t$ precisely characterizes ecological stability in terms of biomass constancy (the denominator) and biomass output (the numerator) of the community's species. Precision is too often lacking when putatively emergent phenomena are discussed, so $S_t$ affords a rare opportunity.

With this account of stability in mind, Maclaurin and Sterelny (2008, section 6.4) argue that it constitutes an emergent property, immune to the reductive craft of MI. Their contrast is labeled an "individualistic" view of communities in which abundances and distributions of organisms are largely controlled by abiotic factors, not inter-organismal interactions. As embodied by seminal figures such as Henry Gleason and Robert Whittaker (see Worster 1994), this view reflects a long-standing theoretical orientation in ecology according to which ostensibly community-level properties simply derive from patterns of abiotic factors impacting different species; lower-level interactions are causally negligible, or fail to produce anything novel at the community level. The marked difference between this austere conception of individualism and MI should be clear. While MI countenances interactions between individuals in accounts of higher-level phenomena, the conception Maclaurin and Sterelny describe corresponds most closely with methodological atomism (see Section 1.2). Such a constraint would seem to preclude canonical cases of past reduction – thermodynamics to statistical mechanics, Newtonian mechanics to relativity theory, and so on – reductions that depend indispensably on lower-level entities *interacting* in specific ways. Championing ecological emergence based on the implausibility of ecological atomism would therefore be a victory too easily achieved. Marginal controversy exists, but as Maclaurin and Sterelny themselves document, few contemporary ecologists doubt biological interactions such as competition, predation, obligate mutualism, and others shape the composition and structure of communities.[9]

Nor is any conceptual difficulty posed for MI by the substantive propositions Maclaurin and Sterlney (2008) associate with emergence, that communities are "real" or that there are causally efficacious properties usually described at the community level.[10] These are open questions that depend on empirical data, and both contentions are consistent with MI. Tilman's own extensive grassland studies have attempted to demonstrate, for instance, that community stability is responsible for and/or produced by diversity, another community-level property (Tilman et al. 2006). And Sterelny (2001) himself convincingly argues for the reality of communities (at least in the past) based on extensive analysis of paleoecological data. These are reasonable contentions, but they do not establish anything meriting the label "emergence," which Maclaurin and Sterelny characterize as the idea that community properties "are not just an extrapolation of" (2008, 113) or are "not simple reflections of" (2008, 120) the properties of individuals. Emergence so conceived, they claim, challenges the reductive view that "all important system-level behavior can be explained, and can only be explained, by explaining the behavior of the parts" (2008, 120).

[9] Note that even thoroughly "individualistic" communities in Maclaurin and Sterelny's sense can be highly temporally stable. Without further restrictions, presumably on interspecific covariances, nothing about $S_t$ prohibits or renders it unlikely that biomass means are high (numerator) and variances and covariances are low (denominator) in communities governed primarily by abiotic factors of the kind described. $S_t$ may therefore poorly indicate the higher-level causal integration they take emergence to require.

[10] "Communities are real, causally important ecological systems if they have emergent or ensemble properties" (Maclaurin and Sterelny 2008, 113).

Before turning to whether temporal stability is emergent in the intended sense, one might think that once causally efficacious properties are described at a higher level, emergence has been established. This criterion is much too weak. Car crashes offer a straightforward example. In analyzing a crash, a car's total mass is (correctly) represented as causally efficacious. It accounts, for instance, for how far a car penetrated a storefront, propelled a pedestrian, and so on. But the total mass is simply the sum of, and therefore simply reflects, the masses of car components. Causal efficacy at a higher level therefore provides a poor guide to emergence. And note that beyond such straightforward cases, the same conclusion holds for more complex higher-level properties. The car's propulsive causal capability, for example, reflects its parts and their interactions, and explanations of that capability would thereby derive from behaviors of its parts.

What is missing from Maclaurin and Sterelny's analysis is an account of how temporal stability is any different. If emergence amounts to anything, it precludes reductive analyses of higher-level properties in terms of lower-level properties. The calculation is obviously more complicated than summing car part masses, but in an explicit formula, Equation (2) indicates exactly how a community-level property reflects population-level properties, namely, biomass means, variances, and covariances. Once population-level details are set, not only is this community-level property fixed, but its precise magnitude can be calculated, and the contributions different population biomasses make to it can be inspected. Population biomasses, in turn, exhibit a much simpler, aggregative relationship with the biomasses of individual organisms comprising them.[11] It therefore seems that temporal stability is not emergent. It can be reductively accounted for by the properties and behavior of a community's parts.

## 1.4 *Reductive Unification in Ecology*

In trade-offs between breadth and depth, MI and reductionism are usually faulted for aligning too rigidly with depth. But scientific history shows that charge is frequently unfair. Finding underlying connections or

[11] In fact, at least two types of IBMs, so-called fixed-radius neighborhood and zone of influence models, have been developed to study how the distribution of sizes of individual organisms, their density in areas, and physiological differences might influence populations' overall biomass (see Grimm and Railsback 2005, section 6.7).

commonality is often the means to greater generality; reduction is sometimes the conduit of unification.

There are several unificatory targets in ecology. One of the most prized is the elusive integration of classical population and community ecology (see Section 1.1) and ecosystem ecology (see Hagen 1989, 1992). Biological entities are the primary focus of the former, and its models address issues concerning species composition, intraspecific dynamics, and interspecific interactions between coexisting species. The abiotic environment has a small, and rarely an explicit, role in these models. In contrast, energy and material flows through biotic interactions and abiotic mechanisms are the principal focus of ecosystem ecology (see Chapin et al. 2011). For instance, tracing cycles of nitrogen as it is processed by organisms and flows through physical channels is one of many such targets of analysis.

Given their contrasting perspectives, models developed in each subfield differ markedly. The causal relations represented are quite dissimilar, and even the variables considered are usually highly disparate. The prospects of synthesizing such divergent modeling approaches therefore seem dim, and, unsurprisingly, relatively few ecologists have attempted to do so. Loreau's (2010) recent and cogent attempt, however, is not only one of the most promising, but it also showcases MI's salience for ecological modeling.

Loreau tries to integrate several models in population-community and ecosystem ecology, which involves making connections between, for example, intra-trophic-level species diversity and ecosystem functioning, interactions in food webs and ecosystem functioning, indirect mutualism and nutrient cycling, population stability and ecosystem stability, and many others. Surveying such extensive and mathematically sophisticated results is impossible here, but fortunately there is a manageable core. The edifice's foundation is a theoretical insight about how population-community and ecosystem perspectives can be bridged that Loreau then applies, elaborates, and scales in different contexts to affect broader integration. And, crucially, the connection concerns the mass and energy budgets *of individual organisms*. Before giving the relevant equation, Loreau explains why such a bridge should exist:

> [E]cosystem ecology and eco-physiology share the concepts of mass and energy budgets as tools for understanding the acquisition, allocation, and disposal of materials and energy in the metabolism and life cycle of both organisms and

> ecosystems. On the other hand, growth and reproduction are the two processes at the individual level that are responsible for population growth, and these processes place high demands on energy and materials in the metabolism of individual organisms. Thus, *the unification of population and ecosystem approaches should be rooted in the ecophysiology of organisms, in particular, in the constraints that govern the acquisition, allocation, and disposal of materials and energy*. (2010, 9; emphasis added)

Loreau believes this equation captures the insight:

$$P = \varepsilon(C - \mu B) > 0, \tag{3}$$

where $P$ designates total organismal production (tissue growth and reproduction); $C$ designates energy consumption; $B$ designates organismal biomass; $\mu$ designates mass-specific basal metabolic rate; and $\varepsilon$ designates production efficiency beyond basal metabolism, that is, the efficiency with which consumed energy is converted into tissue growth and/or reproductive output beyond what survival requires.

If $P > 0$, organisms consume enough energy to grow and/or reproduce. When aggregated across individuals, this determines whether populations grow or shrink. Equation (3) therefore links factors typically in ecosystem ecology's purview (available energy in organisms' environments and organismal physiology) with individual growth and, scaling up, population growth. Through this link, familiar models of population ecology, such as the logistic equation and Equations (1a) and (1b) from above, can incorporate principles of and models developed within ecosystem ecology.

The ingenuity of this individual-level insight is atypical. The two domains of models being reconciled both largely ignore individual variation,[12] and Loreau follows suit to affect their integration. He recognizes, however, that the mathematical "convenience" (2010, 13) of this unrealistic abstraction constitutes a nontrivial limitation. It is one ecologists working with IBMs are unwilling to accept. In fact, Equation (3) and related equations are commonplace within IBMs, and spatially explicit representations of individual-level variability of organismal production often yield significantly different results than when all individuals are idealized as equal

[12] See Roughgarden 2012 for an overview of the exceptions.

(see Grimm and Railsback 2005, section 7). Loreau does not pursue individual-based modeling, but the crucial insight undergirding his expansive integrative project nevertheless stems from the representational import of the individual level.

As a relatively new modeling strategy in ecology, individual-based modeling does not benefit directly from the mathematically sophisticated techniques developed to analyze classical ecological models over several decades. Instead, IBMs opt for enhanced specificity and the greater realism it furnishes by making individual organisms the explanatory and representational priority. In so doing, individual-based modeling draws from a general, methodologically individualistic approach to understanding the world with a long tradition in the social sciences. It is as fruitful in ecology as it is there.

## 2 Potential Perils of Physics Envy

Cross-pollination between disciplines can be fruitful, as the previous section suggests. But it certainly need not be, and occasionally it is not, unfortunately. This section considers a case that reveals the potential perils of appropriating theoretical concepts, results, and techniques from other sciences, physics in this case, when modeling the ecological world.

The case concerns stability. Chapter 3 presented a comprehensive taxonomy of distinct conceptions of ecological stability, and argued that the concepts of resistance, resilience, and tolerance jointly adequately define it. Despite its salience, this complexity is rarely addressed in the extensive literature devoted to mathematical modeling of ecological systems. Instead, ecological stability is typically characterized as Lyapunov stability, a concept of common currency within physics. This characterization has the advantage of integrating the amorphous concept of ecological stability into a well-developed mathematical theory fruitfully utilized in many sciences. Lyapunov stability theory was developed in dynamical systems theory and has been used in applied mathematics to study several subjects, for instance, mechanics, electrical circuits, and economic systems (Hirsch and Smale 1974; Hinrichsen and Pritchard 2005). Lyapunov stability, however, fails to represent the domain of application of ecological stability adequately. This failure illustrates an important limitation of the theory of Lyapunov stability within mathematical ecology.

## 2.1 *Lyapunov Stability*

Lyapunov stability derives its name from the Russian mathematician who first precisely defined the concept to describe the apparently stable equilibrium behavior of the solar system (Lyapunov 1892). His definition has found widespread application outside this context and is frequently used to analyze mathematical models of biological communities. May (1974) used this definition, for instance, in his highly influential analysis of relationships between the stability and complexity of such models (see Justus 2008a).

The definition has some clear advantages. Unlike other definitions, it integrates ecological stability into a thoroughly studied mathematical theory that has proved fruitful in many sciences, especially mathematical physics. It also seems to formalize the important intuition that ecological stability depends on a community's response to perturbation (see Chapter 3 and below). These apparent advantages, however, are specious. Ecological stability is poorly defined as Lyapunov stability.

What exactly is Lyapunov stability? Answering that question requires some background elaboration. As a technical concept developed in a highly mathematical science, its precise content cannot be conveyed in informal terms.[13] Fortunately, the concept's key features can be captured diagrammatically (see Figure 6).

First, the system of interest, in this case a biological community, is represented as a vector $\mathbf{x}(t)$ ($t$ represents time) in a mathematical space called a state space. The components of this vector represent the populations of different species that comprise the community, usually as designated by their abundances. If those abundances change over time, as they almost always do, $\mathbf{x}(t)$ takes different positions in the state space and can exhibit various trajectories and patterns. The equations, usually differential or difference equations, that model the community's dynamics – intrapopulation dynamics such as density dependence, how populations of different species interact (competitively, mutualistically, commensally, etc.), how abiotic factors impact these dynamics and interactions, and so on – determine how those abundances may change. Equations (1a) and (1b) from above, which focus exclusively on predator–prey relations, is a classical example.

[13] See Justus 2008b, sections 2–4, for a concise technical account, and Logofet (1993) for a comprehensive exposition in an ecological context.

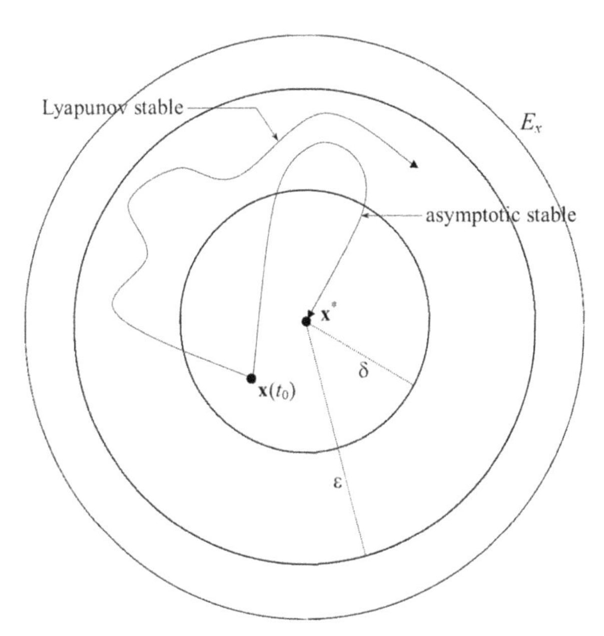


Figure 6 Graphical representation of asymptotic and nonasymptotic Lyapunov stability.

Lyapunov stability concerns how a community behaves in the vicinity of point equilibria, such as $\mathbf{x}^*$ in Figure 6. At a point equilibrium, the abundances of a community remain constant *unless perturbed*. Whether a community has such an equilibrium depends entirely on the equations governing its dynamics. Specifically, setting the left-hand sides of Equations (1a) and (1b) to 0 stipulates that the predator and prey abundances remain constant. Solving the equations then reveals their equilibrium abundances, assuming there are any. Often, particularly when the equations representing community dynamics are more realistic and hence complex, there are none.[14] In the language of Chapter 3, the equations in question constitute the system description *M* and the point equilibrium constitutes the reference state *R* for ascribing Lyapunov stability.

Lyapunov stability is a perturbation-based stability concept: the behavior it requires depends on how the system responds to perturbation. In Figure 6, perturbation is represented by the initial displacement at time $t_0$ from $\mathbf{x}^*$ to $\mathbf{x}(t_0)$.

[14] This poses a nontrivial limitation. Lyapunov stability–like properties can be ascribed to systems exhibiting more complicated reference states or dynamics, but the analytic tools described below, the tools that make the framework so fruitful in so many disciplines, are no longer realizable (see Hinrichsen and Pritchard 2005).

In ecological settings different positions in the space represent different values of the abundances comprising the community, so the perturbation represents any exogenous disturbance that affects those abundances. A drought, severe change in temperature or precipitation, or disease that eradicates portions of some populations are common examples. In this way, Lyapunov stability seems to capture precisely (and formally) the intuition that ecological stability depends on community response to perturbation. Some other popular definitions that define ecological stability in terms of the variability of the populations they contain are mathematically precise in that they are statistical functions of model variables (e.g., Tilman's [1999] "temporal stability"; see Section 1 above). But they do not, as Lyapunov stability does, provide a mathematical characterization of equilibrium system *dynamics*. If $\mathbf{x}^*$ is Lyapunov stable in some region, for instance, its definition ensures predictions can be made about system behavior after perturbation from $\mathbf{x}^*$. Mathematical definitions in the statistical sense cannot ground such predictions unless future values of variables depend on past values in some systematic way, and the precise form of the dependency is known.

Put simplistically, but digestibly, Lyapunov stability requires something that resembles the claim that the effects of perturbations are circumscribed. Consider the region within $\varepsilon$ of the equilibrium $\mathbf{x}^*$ (i.e., the $\varepsilon$-neighborhood around $\mathbf{x}^*$). Lyapunov stability dictates that for *any* such region surrounding $\mathbf{x}^*$, a smaller, interior region within $\delta$ of $\mathbf{x}^*$ *always exists* such that if the system is perturbed within that smaller region (i.e., the $\delta$-neighborhood around $\mathbf{x}^*$), it forever remains within the larger $\varepsilon$ region.[15] $\mathbf{x}^*$ is *asymptotically Lyapunov stable* if in addition to exhibiting this behavior it also tends to return to $\mathbf{x}^*$ (and in the limit does so). The space within which systems are (asymptotically) Lyapunov stable ($E_x$ in Figure 6) is called the *(attraction) stability domain* of $\mathbf{x}^*$ (see the discussion of tolerance in Chapter 3).

The formal precision of this definition unfortunately does not help circumvent a serious difficulty: directly assessing whether a system would appropriately respond to perturbation is usually infeasible because explicit

[15] For formal purists, an equilibrium $\mathbf{x}^*$ is *Lyapunov stable* iff: $(\forall\varepsilon > 0)(\exists\delta > 0)[|\mathbf{x}(t_0) - \mathbf{x}^*| < \delta \Rightarrow (\forall t \geq t_0)(|\mathbf{x}(t) - \mathbf{x}^*| < \varepsilon)]$, where $\varepsilon$ and $\delta$ are real values ($\delta \leq \varepsilon$). In textbooks, Lyapunov stability is frequently glossed as saying that the effects of perturbation are always bounded (see Kot 2001, 9). But that gloss is misleading. Interpreted straightforwardly, it inverts the first two quantifiers, making the definition begin $(\forall\delta > 0)(\exists\varepsilon > 0)$. Thanks to Peter Vranas for stressing this incongruity.

solutions for $\mathbf{x}(t)$ are rare. The systems scientists study, biological communities being no exception, are typically quite complicated. The differential equations representing them are no different. Faced with such complexity, analytic solution techniques have serious limitations. Those limitations in turn reveal serious challenges to actually evaluating the system behavior the definition of Lyapunov stability requires.[16]

Lyapunov (1892) himself recognized this difficulty and developed two methods for assessing Lyapunov stability indirectly. The first "indirect method" involves "linearizing" the differential questions at the point equilibrium $\mathbf{x}^*$. In effect, the differential equations, which may be highly nonlinear, are approximated with linear equations. Since this linearization is defined only at $\mathbf{x}^*$, stability determined by the indirect method is restricted to *infinitesimal* neighborhoods of $\mathbf{x}^*$. It is accordingly called *local stability*. The indirect method is prevalent in mathematical modeling because it is almost universally applicable: it can be applied to any system representable by continuous and continuously differential equations (Hinrichsen and Pritchard 2005), which is the norm in ecology.

Although the frequent focus of community modeling, perhaps because it is relatively easy to evaluate with the indirect method (Hastings 1988), local stability characterizes ecological stability extremely poorly. Differentiable models of systems can be linearized at any point of state space, such as at an equilibrium. By itself, however, linearization provides information only about local system behavior, that is, only in an *infinitesimal* neighborhood of the linearization point. Local stability therefore says nothing about system behavior outside this extremely restricted domain, and stability criteria based on the indirect method therefore obviously provide no information about the extent of attraction or stability domains. Besides it being fundamentally unclear how perturbation within an infinitesimal domain can be biologically interpreted (or empirically measured), real-world perturbations are clearly not of infinitesimal magnitude. Any real perturbation will expel a system at a strictly locally stable equilibrium from its stability domain. Local stability of nonlinear systems like biological communities consequently provides no information about their response to common perturbations like drought, fire, population reduction by disease, and so on. Application of the indirect method to community models, which can evaluate only local

[16] Recall the universal temporal quantification in the formal definition: $(\forall t \geq t_0)$.

stability, therefore yields negligible insight into the community dynamics relevant to ecological stability. For this reason, food web modeling of community stability, which is based on the indirect method, has been trenchantly criticized (Hastings 1988). This is a clear case of mathematical expediency coming at the expense of ecological relevance.

This kind of deficiency prompted Lyapunov (1892) to develop a "direct method" for evaluating stability that does not rely on linearization. It involves constructing a specific function, called a Lyapunov function, with an origin at $\mathbf{x}^*$ (see Justus 2008b for details). The existence of such a function is provably necessary and sufficient for $\mathbf{x}^*$ to be stable. As a methodology, it is called "direct" because whether such a function can be constructed depends directly on the mathematical form of the equations representing a system, unlike the indirect method, which relies on their linearization, but unlike direct evaluation of perturbation response behavior, which requires explicit solutions for $\mathbf{x}(t)$.

Although no general method for constructing Lyapunov functions is known, the direct method has proven to be an extremely useful tool for analyzing physical systems, especially in the classical framework governed by only Newtonian mechanics and friction. Across scientific fields, this is exceptional rather than typical. Constructing Lyapunov functions is usually extremely difficult for the predominantly nonlinear systems that scientists study (Goh 1977). The reason for its utility in the classical framework is twofold. First, there are highly confirmed mathematical models describing numerous types of systems in this framework. The likelihood is therefore high that application of the direct method to these models will reveal the true stability properties of the systems they accurately represent. Second, there are certain quantities, such as total energy, that are conserved or monotonically dissipated in such systems, depending on how they are characterized (closed or open). These quantities ensure Lyapunov functions exist for models of these systems. Lyapunov, in fact, developed the Lyapunov function to generalize the classical energy concept, and energy conservation played an important role in his development of the proof about the connection between stability and Lyapunov functions (Lyapunov 1892).

Within the classical framework, stability properties of more complicated system models that include and disregard friction can be evaluated with the direct method. The method is also useful in nonclassical frameworks with similar properties, such as mass-energy conservation in special relativity theory. The equations characterizing systems in these and the classical framework are often too complex to solve analytically, and the direct method provides the only means by which attraction or stability domains of equilibria can be determined. In a wide variety of frameworks within physics, therefore, the direct method and concept of Lyapunov stability are indispensable. And given that the direct method adequately characterizes the equilibrium dynamics of many systems in physics (and other sciences), it seems reasonable to expect the method will function similarly for ecological systems since their models share a similar mathematical structure. Like systems studied in physics, biological communities are usually modeled by differential equations. The only difference is that components of the vector $\mathbf{x}(t) = \langle x_1(t), \ldots, x_i(t), \ldots, x_n(t)\rangle$ represent biological variables, usually population sizes of species in a community, instead of typical physical ones. The structural similarity between the two domains has been a persistent thread in theoretical ecology, from the work of early ecologists such as Alfred Lotka and Vito Volterra (see Kingsland 1995) to the physicist-turned-ecologist Robert May, the mathematician-turned-ecologist Robert MacArthur, and the legions of their students and other students of the prolific G. E. Hutchinson. If it's good enough for physics, surely it's good enough for ecology.

Before dashing these hopes, yet another apparent advantage should be highlighted: properties of community response to perturbation that are fundamental components of ecological stability à la the analysis in Chapter 3 *seem* to be formalizable with the direct method. The size of attraction domains of asymptotically Lyapunov stable equilibria and the rate that which systems return to them, for instance, are definable in terms of Lyapunov functions. The larger the attraction domain ($E_x$ in Figure 6), for instance, the stronger the perturbations of variables a system can sustain and return to $\mathbf{x}^*$, which is often called system "tolerance." Similarly, the "steeper" the Lyapunov function, a quantity that can be carefully characterized mathematically, the faster it returns to $\mathbf{x}^*$, often called system "resilience." If Lyapunov stability adequately defined ecological stability, it would therefore be possible to formalize the properties of tolerance and resilience and their relation to the stability of biological communities. Specifically, the intuitive idea that ecological stability increases with the strength of perturbation from which a system can return to equilibrium and with the speed of return to equilibrium after perturbation would be formalizable.

Now for the bad news. Despite its apparent advantages, interpreting Lyapunov stability in ecological terms furnishes strong reasons for rejecting the definition. In general, the adequacy of definitions depends on what they require of defined concepts. Characterizing ecological stability as Lyapunov stability stipulates conditions that biological communities must satisfy to be ecologically stable. If these conditions are unreasonable – if they are biologically unrealistic, too weak, too strong, or otherwise problematic – the definition is inadequate. In this case, the antecedent is unfortunately veridical.

To see why, and given that local Lyapunov stability is woefully too weak as a definitional candidate, consider characterizing ecological stability as the strongest Lyapunov stability property: *global* asymptotic stability. A system with a globally asymptotic stable equilibrium will return to it following *any* perturbation from it (provided the perturbation does not destroy the system). That is, $E_x$ in Figure 6 would span the entire state space. This certainly appears to be much stronger than ecological stability: it seems to entail that a community would return to equilibrium after any perturbation that does not eradicate its species. Model variables of systems at asymptotically and globally stable equilibria that are perturbed to 0.01% of their equilibrium values, for instance, will deterministically return to their initial values. Although this response to such severe perturbations is not required for a biological community to be ecologically stable, it certainly seems sufficient.

But this assessment is mistaken. It is based on an indefensibly narrow conception of ecological perturbation. Real-world perturbations to biological communities do not merely affect population abundances, represented by *variables* in models like Equations (1a) and (1b) above. They also temporarily change environmental factors that influence community dynamics, represented by model *parameters*. As May (1974, 216) creatively put it:

> in nature, population perturbations are driven not by the stroke of a mathematician's pen resetting initial conditions [i.e., resetting variable values to represent perturbations], but by fluctuations and changes in environmental parameters such as birth rates, carrying capacities, and so on.[17]

[17] This remarks occurs in May's (1974) "Afterthoughts," in which he goes on to suggest that a fully adequate account of the property labeled "tolerance" from chapter 3 will require an assessment of how community dynamics changes as environmental parameters change. In spite of this disclaimer, the predominant focus of his 1974 book, numerous subsequent articles, and later mathematical modeling concerned with ecological stability in general has been Lyapunov stability.

Perturbations may induce, for instance, changes in environmental parameters that modify the strength and qualitative nature of interspecific and intraspecific community interactions. Initially noninteracting species may begin competing or exhibiting other nonneutral interactions, and vice versa. Since real-world perturbations may affect environmental factors that influence community dynamics as well as population sizes of species, they should be represented by corresponding changes in variable *and* parameter values. Ecological stability should then be evaluated with respect to system behavior following the full suite of possible types of changes, perturbations represented as temporary changes to variables ($P_v$), perturbations represented as temporary changes to parameters ($P_p$), and perturbations that should be represented as temporary changes to both ($P_{vp}$) (see Chapter 3).

Lyapunov stability, however, only assesses system response to temporary changes in *variable* values. It does not consider system responses to changes in parameters and thus provides only a partial account of the type of system response to perturbation that ecological stability requires. Lyapunov's direct method is therefore not a suitable methodology for analyzing ecologically relevant nonlocal stability properties of standard community models, although it accurately estimates stability to perturbations affecting only model variables.

This generates problematic (and sometimes quite bizarre) judgments. For example, systems can simultaneously have globally asymptotically stable equilibria, but also be exceedingly structurally unstable, that is, constituted such that very slight parameter changes (representing insignificant perturbations) can produce complete system collapse (see Hallam 1986). In other contexts, global asymptotic stability is strictly dependent on there being an even number of system components; that is, only communities with even numbers of species could exhibit this stability in such cases.[18] The defect may lie in the specific differential equations used to represent a community, but surely a more adequate characterization of ecological stability would have offered greater skepticism about a connection between evenness of species richness and stability.

[18] It is possible, of course, that such a relationship exists but has not yet been found; and most ecologists highly doubt it. May (1974, 53), for instance, suggested that it "border[s] on the ridiculous," and Volterra, who first recognized the fact, was similarly skeptical (Scudo 1971).

In general, Lyapunov stability inadequately captures the conviction that ecological stability depends on community response to perturbation because it inadequately represents real-world perturbations. Whereas Lyapunov stability focuses strictly on system behavior in *state* space, ecological stability essentially concerns system behavior in *parameter* space as well. Consequently, Lyapunov stability is not a sufficient condition for ecological stability, as some ecologists have explicitly suggested (e.g., Logofet 1993, 109) and as many ecologists assume.

That ecological stability should not be defined as Lyapunov stability does not entail that the latter should play no role in mathematical modeling of biological communities. For a community at equilibrium, asymptotic Lyapunov stability within nonlocal attraction domains adequately represents ecologically stable responses to perturbations that alter species population sizes but leave community structure and parameter values unchanged. Lyapunov stability is therefore a plausible necessary condition for ecological stability. Moderate culling of some community species or temporary cessation of plant or animal harvesting in ecosystems where it has a long history are plausible examples of such perturbations. In fact, the effects that fishing cessation during World War I would have on fish populations motivated Volterra to begin developing mathematical models of communities (Scudo 1971).

This analysis also does not detract from the obvious gains in precision, rigor, and other scientific virtues afforded by mathematical dynamic systems modeling within ecology. It does expose, however, a limitation of the stability definition that usually accompanies this type of modeling. If ecological systems like communities were completely analogous to systems studied within physics for which Lyapunov's methods are so successfully applied, defining ecological stability as Lyapunov stability would be entirely justifiable. As we have made clear, however, there is a crucial difference in how perturbations should be represented in the two contexts. Scientific definitions that make an initially problematic concept precise by integrating it into a systematically developed mathematical theory, even one fruitfully utilized in other sciences, are therefore not always satisfactory.

Within classical mechanics, force fields governing interactions between bodies, such as gravitation, are usually invariant. Masses of bodies in the system, overall system structure (e.g. mass–spring system, damped pendulum, mass–pulley system), and other system features are also usually held or assumed fixed in this framework. Model features representing these system properties, such as parameter values and model structure, are correspondingly fixed. With the background held constant in this way, perturbations are represented by temporary finite changes in variable values, for instance, by displacements of position vectors and alterations to velocity vectors. Energy is conserved or monotonically dissipated in these systems depending on whether they are closed or open, so Lyapunov functions can always be constructed for them and their stability properties evaluated. Since perturbations of such systems are usually not taken to change the system properties represented by parameters and model structure, Lyapunov stability captures their important stability properties.

In ecology, however, the underlying facts are very different. Parameters representing external factors affecting species and their interactions are much more likely to change than in systems studied within classical mechanics because they are regularly altered by real-world perturbations.[19] An appropriate definition of ecological stability therefore requires integrating Lyapunov stability with a concept representing how communities respond to these types of structural change: a structural stability concept. This would adequately account for how communities respond to real-world perturbations represented by temporary changes in variable *and* parameter values. Resilience and tolerance to real-world perturbations would be representable with such a concept.

This suggestion poses a formidable challenge to mathematical ecologists because structural stability is a much more technically complex mathematical concept than Lyapunov stability (see, e.g., Peixoto 1959). The dearth of work, especially biologically oriented work, devoted to it (Lewontin 1969; May 1974), however, may explain its lack of application within mathematical ecology, not any essential mathematical intractability of the concept.

The first prominent mathematical ecologists, Lotka and Volterra, were physicists by training, and this significantly influenced their approach to modeling biological systems (Kingsland 1995). Their work, moreover, subsequently set much of the agenda of twentieth-century mathematical ecology. Not surprisingly, most mathematical ecologists have used the concept of Lyapunov stability and the direct and indirect methods to analyze community models. But historical inertia does not a justification make.

[19] The difference between ecological systems and those typically studied in physics should not be overstated. In fluid mechanics, for instance, the background structure of the systems studied is highly variable. This analysis suggests, therefore, that Lyapunov stability would not be an appropriate representation of the stability of these systems.

# 5 Biodiversity

Some ecological systems are more complicated than others. For example, tropical communities usually contain more species (Pianka 1966; Willig et al. 2003); there is evidence their species interact more intensely (Janzen 1970; Møller 1998); these interactions are more variegated in form (Dyer and Coley 2001); and they exhibit more trophic levels than high-latitude communities (Oksanen et al. 1981; Fretwell 1987). Ecologists often use the concept of diversity to represent differences in the "complicatedness" of such systems. For example, tropical communities are often said to be more diverse than tundra communities.

At a coarse level of description, the vague connotations accompanying the term "diversity" adequately reflect the imprecise judgments that some systems are more complex or complicated than others. Disagreement arises, however, over how the concept should be defined and, in turn, operationalized, or indeed whether any such characterization is possible. As early as 1969, Eberhardt (1969, 503) deemed the ecological literature on diversity a "considerable confusion of concepts, definitions, models, and measures (or indices)." A few years later, Hurlbert (1971, 577) argued that "the term 'species diversity' has been defined in such various and disparate ways that it now conveys no information other than 'something to do with community structure.'" MacArthur (1972, 197) similarly suggested that the term "diversity" should be excised from ecological vocabulary as doing more harm than good and that ecologists had "wasted a great deal of time in polemics about whether [Simpson's] or [Shannon's] or $\frac{N!}{N_1!N_2!\ldots N_n!}$ or some other measure [of diversity] is 'best.'" As these remarks indicate, ecologists had proposed several precise measures of diversity that differed not only in mathematical form but also, and more importantly, in what properties are given priority over others. Disagreements about these issues had therefore broached the question of which properties of ecological systems *should* be considered part of their diversity and, in turn, which features an adequate characterization of the concept should possess. Long before the term "biodiversity" ever emerged, "diversity" was a problematically complex concept in ecology.

Compared with the contemporary confusion, controversy, uncertainty, and profound ethical significance surrounding the concept of biodiversity, the earlier disagreements among ecologists appear rather mundane. Besides its interest to ecologists, the significance of biodiversity is now common currency in environmental ethics and at the center of much broader debates about species loss, the value of the natural world, and how humans will have to rethink their relations to that world to save it. As a concept that both depends on and can contribute to those debates, biodiversity raises especially interesting philosophical issues about the sciences studying and endeavoring to protect it, and in turn presents an interesting case study about how seemingly narrow conceptual issues can intertwine with fundamental and wide-ranging theoretical, methodological, epistemological, and ethical issues within science and society in general.

## 1 Explicitly Defining Diversity

Coined as a simple shorthand for "biological diversity" by Walter Rosen for a conference "National Forum on BioDiversity" in 1986 (Wilson 1988; see Takacs 1996), at its most inclusive "biodiversity" designates diversity at all "levels" of biological organization. In ecology, the most common focus is the population and community levels, but developmental, genetic, morphological, phylogenetic-taxonomic, and even geographical considerations are arguably in its purview (Maclaurin and Sterelny 2008). So, at a minimum, it seems biodiversity spans species richness; relative abundances; genetic, phenotypic, and phylogenetic diversity; functional diversity; and perhaps differences in what is represented across geographical areas (often called "complementarity"), and more.

This expansive generality is problematic. Defining biodiversity as "all that" is obviously pointless. Capturing full generality results in nothing informative. But that does not mean definitions for *portions* of biodiversity are impossible or necessarily uninformative. How this multifaceted diversity, and portions thereof, should be characterized crucially depends on how biological systems are represented, particularly on how their parts are individuated, classified, and distributed among those classes. And representations may vary with different explanatory or predictive scientific goals, and across types of systems, so characterizations of diversity may vary across these contexts as well. As it so often does, in this case it pays to start simple.

In ecological contexts, ascriptions of diversity typically target biological communities. Like most systems studied in science, biological communities can be represented with different degrees of detail. With low specificity, a community can simply be described in terms of the species it contains and how individual organisms of the community are distributed among these species. The proportional species abundance vector captures this information:

$$V_p = \langle p_1, \ldots p_i, \ldots, p_n \rangle,$$

in which $n$ designates the number of species in the community, called its *species richness*; $p_i$ designates the proportional abundance of the $i$th species in the community; the $p_i$ are ordered from most to least abundant; and, of course, they sum to 1. The only properties of species $V_p$ represents are proportional abundances. Functional, morphological, trophic, and taxonomic differences (beyond the species level) are ignored. Proportional abundances of species in a community often change over time for a variety of reasons (e.g., migration, interspecific interactions such as predation, competition), so $V_p$ must be updated as communities evolve.

Each $p_i > 0$. That is, to be one of the species comprising a community, the community must contain at least one representative of that species. As a biological collection, to deny this stipulation for communities would require commitment to the idea that a community can be represented to contain a species not instantiated by any of its members. In modeling contexts with different goals, such as in studies of extinction and migration processes, it may be useful to allow zero $V_p$ components to represent when species have gone locally extinct or have emigrated completely from a community. Once individuals of a species disappear from a community, however, that species is no longer part of that community and does not contribute to its diversity.

"Abundance" in "proportional abundance" is ambiguous. Besides referring to the number of individual organisms of a species (a discrete quantity), it can also refer to their biomass (a continuous quantity). Accordingly, $p_i$ can designate either (1) the proportion of individuals of species $i$ in a community ($N_i/N$) where $N_i$ is the number of individuals of species $i$ and $N$ is the total number of community individuals, such as the proportion of wolves in an community, or (2) species $i$'s proportion of total community biomass, such as dry weight of a particular plant species in a forest community. $p_i$ may differ significantly on these two interpretations, so ideally $V_p$ should be calculated according to both interpretations for a given biological community. If it is unclear how to count individual organisms, as is the case for some clonal plant species or asexually reproducing marine species, the biomass interpretation of $p_i$ is preferable.

If only this minimal information is considered, ecologists widely agree that two properties determine a community's diversity: *species richness* and *evenness* (Margurann 2004). Consider an example, two hypothetical simple communities, A and B, both composed of two species $s_1$ and $s_2$. A and B have the same species richness (two). If the proportions of individuals distributed among the two species are $p_1 = 0.02\%$ and $p_2 = 99.98\%$ for A and $p_1 = 50\%$ and $p_2 = 50\%$ for B, B is more even than A.

The widespread belief that species richness and evenness are components of diversity reflects the intuitive ideas that diversity increases as either increases. Consider each in turn. That diversity increases as richness increases captures an incontestable feature of the concept: the diversity of a collection increases as the number of different types of entities in the collection increases. For a biological collection such as a community, a community composed of 1,000 species is unquestionably more diverse than one composed of ten.

There is a similar rationale for the relationship between evenness and diversity. Consider two communities, each composed of 100 species and 10,000 total individual organisms. A community in which there are 100 individuals of each species seems more diverse than one with 9,901 individuals of one species and one individual each of the other 99. The reason seems to be that besides a consideration of the number of types of entities in a collection, diversity also involves a consideration of how well they are represented. For this reason, diversity is often interpreted as the *apparent* or *effective* number of species present in a community (e.g., Peet 1974). For example, to an observer with imperfect faculties of perception, or an ecologist with insufficient field time or employing sampling methods with unavoidable limitations, the first community with evenly distributed individuals will usually appear to contain more species than the second community, despite their identical richness. This captures the intuitive idea that community B is more diverse than community A from above.

This rationale can be given a set-theoretic gloss. A biological community can be represented as a set of organisms of different species. Sets are characterized by properties of their members. Members of an uneven community poorly represent some species, while each species of an even community is equally represented by its members. As a set, the characterization of a community with evenly distributed individuals therefore depends more significantly on a greater number of species types than an uneven community.

Compared with richness, the contribution that evenness makes to diversity is more difficult to gauge. $V_p$ gives a determinate specification for species richness ($n$). It does not for evenness. That makes assessing when evenness increases (or decreases) unclear.[1] It is clear, however, when evenness is maximal. For a given species richness, evenness is maximized when individuals of the community are equally distributed among species, that is, when $p_i = 1/n$ for all $i$. Similarly, evenness is clearly minimized when community individuals are maximally unequally distributed. Specifically, it is minimal when all but $(n - 1)$ of the individual organisms comprising the community are of one species and the rest are equally distributed (one each) among the other $(n - 1)$ species. In this way, definite bounds for evenness can be identified.

That diversity increases with increases with species richness and evenness and that there are extrema for evenness is well known (see Lande 1996; Margurann 2004; Sarkar 2007). A key question is how evenness should be evaluated between these extrema, and how it contributes to overall diversity. In 2011, I proposed an answer to this question (Justus 2011). Label the maximally even and minimally even proportional abundances $V_p^{\max}$ and $V_p^{\min}$. Since evenness is maximal for $V_p^{\max}$, evenness must decrease as $V_p$ diverges from it. This decrease can be quantified in many ways, but one defensible rationale for doing so restricts the range of possible methods of quantification. Recall that the only differences between species being considered are their proportional abundances; $V_p$ does not represent taxonomic, trophic, functional, and other interspecific differences. Besides their proportional abundances, different species are therefore treated as equally important in assessing the diversity of a community. Thus, if evenness decreases

[1] In addition, increases in species richness (represented by new $p_i$) necessitate changes in the $p_i$ comprising $V_p$ since all the $p_i$ must sum to one following any change in species richness. These changes do not *necessitate* a change in evenness, but the absence of a determinate specification from $V_p$ makes it unclear how evenness can remain static as species richness changes.

because one species deviates from its maximally even proportional abundance ($1/n$), an equal deviation from the maximally even proportional abundance by another species should induce an identical decrease in evenness and thus diversity for a given species richness.

Equal decreases in evenness for equal deviations from maximal evenness across species ensures that diversity is blind to species identity *when only* $V_p$ *is considered*. It thereby captures the frequently made assumption that evaluating community diversity requires treating species as equals *in the absence of taxonomic, functional, or other data* (Magurran 2004, 11). In such cases, only the extent a species' proportional abundance deviates, not what species it is, is relevant when assessing a community's diversity. In fact, the only way species are individuated via $V_p$ is through their proportional abundances.

The task of showing how evenness should be measured remains unfinished. Being blind to species identity as above requires merely that equal changes in the abundances of two species from maximal evenness necessitate equal decreases in diversity for a given species richness. But such blindness is nevertheless neutral about whether rare or abundant species are more important to the diversity of a community. In particular, different types of deviations from $V_p^{\max}$, such as those involving rare, abundant, or different numbers of species, might be accorded different import for diversity. Of course, unequal deviations of the *same type* should necessitate different values of diversity. Consider, for example, the type of deviation in which one species $i$ deviates from $1/n$. If $p_i$ decreases from $1/n$ to $1/2n$, diversity should decrease less than if $p_i$ decreases to $1/3n$ all else being equal because the decrease in evenness is greater in the latter case. If $p_i$ decreases from $1/n$ to $1/4n$, however, it is as yet unclear whether diversity should decrease more or less than in a case in which $p_i$ decreases from $1/n$ to $1/2n$ and some other $p_k$ decreases from $1/n$ to $1/2n$.

What is needed is a defensible method for evaluating evenness that adjudicates between different types of deviations from $V_p^{\max}$ for a given species richness. And there is such a method: evaluate evenness in terms of the distance between $V_p$ and $V_p^{\max}$ (see Justus 2011 for details). Since components of these vectors take real values, a typical Euclidean distance metric is appropriate. Measured in this way, the idea is that evenness is inversely related to the Euclidean distance between its species abundance vector $V_p$ and $V_p^{\max}$. For example, if the diversity of $V_p^{\max}$ and $V_p^{\min}$ for a given richness are set at 1 and 0, respectively, the diversity of a community represented by

$V_p$ would take values on [0, 1] determined by the Euclidean distance between $V_p$ and $V_p^{max}$. Species abundance vectors that deviate from $V_p^{max}$ in different ways but at the same distance from it would thereby have the same evenness; those at different distances from $V_p^{max}$ would differ in evenness. In particular, for the two species abundance vectors discussed above, one in which $p_i$ for one species decreases from $1/n$ to $1/4n$ and another in which $p_j$ and $p_k$ for two species decrease from $1/n$ to $1/2n$, the latter would be accorded greater evenness because its distance from $V_p^{max}$ is smaller than the former.

Similar to blindness to species identity, measuring evenness by Euclidean distance from $V_p^{max}$ is blind to the type of deviation from $V_p^{max}$. This imposes a significant constraint on the diversity concept. It requires treating changes in the proportional abundances of rare and common species as equally important to diversity. For instance, diversity must decrease the same amount with a decrease in $p_i$ for an extremely rare species and with an identical decrease in a much more common species. This precludes diversity from being partially sensitive to the proportional abundances of rare or common species. In particular, it requires that species abundance vectors in which several species are very abundant and a few are very rare, and in which several species are very rare and a few are very abundant, have the same diversity if their distances from $V_p^{max}$ are identical.

For the same reason that diversity should be blind to species identity and blind to the type of deviation from $V_p^{max}$, diversity should not be partial to particular distances between $V_p$ and $V_p^{max}$ in the sense that diversity should decrease uniformly (i.e., linearly) as the distance between $V_p$ and $V_p^{max}$ increases. This impartiality captures an idea sometimes mentioned in discussions of ecological diversity, that diversity should not be partial among individual organisms, just as it should not be partial among species in a community (e.g. Krebs 1989; Magurran 2004). Specifically, in assessments of diversity in the absence of taxonomic, functional, and other types of information, individual organisms should contribute to diversity in proportion only to the proportional abundance of the species to which they belong. If different types of deviations from $V_p^{max}$ are weighted differently than as dictated by Euclidean distance, some individuals will contribute more (or less) to diversity merely because they are a member of a species that has deviated from $1/n$ in a favored (or disfavored) way. Similarly, if different distances from $V_p^{max}$ are weighted differently in assessing diversity, some individuals will contribute more (or less) to diversity merely because they are a member of a species with a proportional abundance at a distance from $1/n$ that is favored (or disfavored). In either case, individual organisms would not be treated as equals in determining the diversity of a community composed of them.

There *can*, of course, be reasons to treat individuals of different species differently, even in the absence of genetic, phenotypic, phylogentic, or any other information. Individuals of rare species (and their proportional abundances) are usually weighted more significantly in assessing the diversity of communities in conservation biology, for instance. Rare species are usually more likely to go extinct. Assessing diversity such that changes in their proportional abundances are accorded more import than changes in common species is typical in conservation contexts because changes in the former are more likely to influence species persistence than changes in the latter. In such cases, human values rather than anything biological prompt treating species unequally. See Sections 2 and 3 below for an analysis of this approach to characterizing diversity.

Commitment to diversity being species-blind and individual-blind necessitates the specific method of assessing evenness with the Euclidean distance metric described above (for a given species richness). The method therefore has a particularly solid foundation: deviate from it on pain of treating species or individuals as unequal *in the absence of any biological information that would justify such treatment*. The method does not speak, however, to what weight richness and evenness should have in judging diversity. Frankly, I am entirely unclear whether anything determinant and defensible can be said on that issue, beyond banalities such as that both should be considered. They are, after all, logically and conceptually independent properties. Some analyses make more specific proposals. One adequacy condition for diversity proposed by Lewontin (1972) and endorsed by Lande (1996) concerns the relationship between the diversity of individual communities and the diversity of sets of different communities. Specifically, if a super-collection of individuals is formed by pooling the individuals of several distinct smaller collections, the idea is that the diversity of the super-collection must be at least as great as the average diversity of the smaller collections. Applied to biological communities, this requires the diversity of the super-community **C** formed by pooling the individuals of each community $C_i$ to be greater than or equal to the weighted mean diversity of the $C_i$.

Lewontin did not provide a rationale for this constraint on ecological diversity, and there are reasons to reject it as an adequacy condition.

Consider two simple communities $C$ and $D$ composed of four different species (two each) with absolute (not proportional) abundances $\langle 2, 2 \rangle$ and $\langle 1000, 1000 \rangle$. The absolute species abundance vector for the super community $C \cup D$ with species richness four is $\langle 1000, 1000, 2, 2 \rangle$. The proposed adequacy condition requires the diversity of $C \cup D$ be greater than the average diversity of $C$ and $D$, but it is unclear why this is defensible as an adequacy condition on the concept of diversity. $C \cup D$ contains more species than either $C$ or $D$, and in this respect seems more diverse. But it is also highly uneven compared with $C$ or $D$. The proportional species abundance vector for $C \cup D$ is approximately $\langle 0.499, 0.499, 0.001, 0.001 \rangle$, which is a highly uneven distribution, unlike the highly even distribution of $C$ and $D$, $\langle 0.5, 0.5 \rangle$. The condition therefore forces a strong rank order of species richness over evenness in assessments of diversity.

This *may* be a defensible property of a proposed diversity index. In conservation biology, for example, there may be human value–laden advantages to prioritizing species richness over evenness in assessments of the diversity of communities targeted for conservation. The condition is not, however, a defensible constraint on *any* potential quantitative specification of diversity. Species richness and evenness are independent properties, and although this does not entail that one is not more important than another in evaluations of a community's diversity, nothing about the pre-theoretic concept of ecological diversity seems to suggest otherwise. Pielou (1977, 292), for instance, explicitly rejected the constraint: "since diversity depends on two independent properties of a collection ... a collection with few species and high evenness could have the same diversity as another collection with many species and low evenness."[2]

## 2 Value-Driven Explicit Definition

As the term "biodiversity" emerged and the general concept was gaining philosophical and scientific attention in the late 1980s, Bryan Norton, a panelist at the "National Forum on BioDiversity" (Norton 1988) and the only philosopher there, was developing a novel anthropocentric theory of its

[2] In passing, Lande (1996, 8) motivates this condition by pointing out that its denial "implies the possibility of a negative diversity among communities." But why this is problematic is unclear. It does not, for instance, entail that the diversity of any individual community is negative, which would clearly be problematic.

value (Norton 1987). Anthropocentric, that is, human-centered or human-focused, approaches to ethical value, and the value of the natural world in particular, have an enduring and seemingly entrenched standing in human history, at least within so-called Western civilizations. With very rare exceptions, Judeo-Christian traditions place humans, men specifically, at the absolute center of nondivine reality, everything else having secondary and subordinate value (see White 1967). Protagoras' *homo mensura* – "man is the measure of all things" – arguably expresses the same evaluative orientation in a very different cultural context (Kerford 1981). As is so often the case, philosophy reflects history: humans and their interactions are the principal focus of most contemporary ethical theories. Theories with quite divergent commitments – various varieties of utilitarianism, virtue ethics, contractarianism, and deontology – are all broadly anthropocentric. Nonhuman natural entities typically possess value only derivatively and in relation to humans. The desire to counter this view was at the heart of John Muir's famed debate with the then first chief of the US Forest Service, Gifford Pinchot, over the building of the Hetch Hetchy dam (Righter 2005), and it continues to divide conservationists and environmental ethicists today (Justus et al. 2008a).

There is an obvious shortcoming of standard anthropocentric approaches. If portions, and they are probably *vast* portions, of the natural world are insufficiently economically valuable, although ecologically functional and/or aesthetically pleasing to humans (and perhaps a small set of other cognitively sophisticated creatures), those portions simply are judged to be "not valuable." The dismal ethical fortune of so much of the natural world on this approach seems to count definitively against it. Norton's theory is unabashedly anthropocentric, but with an ingenious maneuver it offers the enticing prospects that this environmentalist catastrophe can be avoided without resorting to non-anthropocentric value claims many find highly problematic.

The key was recognizing that the category "valued by humans" as gauged by the explicit and directly assessable kinds of values considered above (economic, ecological, aesthetic, etc.) does not exhaust the category "valuable to humans." Some things and experiences of those things seem to have *transformative* value, that is, the ability to transform humans' direct, first-order desires, preferences, and value judgments. This ability is usually not assessable via these first-order valuations. In an especially compelling example, Norton (1987) described how experiencing classical music concert might have such a transformative power. Absent any experience of this music beforehand, a concert ticket might have no demand value for someone, or even negative value ("what a boring waste of time"). But actually experiencing the music might, and it seems frequently does, change those first-order demand values in a positive direction. Norton's thesis, pursued in several cogent analyses rich in well-crafted and insightful examples (see Norton 1994, 2005), is that experiences of the natural world, biodiversity in particular, might be similarly transformative.

The significant appeal, sophistication, and serious challenges to Norton's (1987) ethical theory fall outside the purview of this analysis (see Sarkar 2005, chapter 4; Odenbaugh 2009). But as Norton presciently alluded to (1987, chapters 12 and 13)[3] and later made explicit (2006), he believed his account of the value of biodiversity bears directly on the task of defining it, our present concern.

The project, ironically, is to winnow diversity when defining biodiversity. The simplest, most comprehensive extension of "biodiversity" includes diversity at *all* levels of biological organization, from expansive ecosystems at the broadest scale all the way, presumably, to cellular organelles or smaller. But characterizations of the concept that directly and exhaustively mirror that diversity are effectively useless. To focus on just one "level": significant genotypic and phenotypic differences exist across members of the same species; this diversity is in fact the necessary grist for natural selection. But the implication that the cause of biodiversity conservation is advanced by deploying limited funds to protect this diversity in toto in, say, North American white-tailed deer

[3] "Where, then, does the priorities issue stand? An assessment may suggest some new directions. First, it would appear that formal priority criteria will cause little controversy so far as they go. Formal criteria merely recognize and implement the general value of diversity. Any goal, including the one of preserving species, should be undertaken efficiently. Efforts should be expended on species that are in trouble but salvageable and on species that can be saved inexpensively. Within this context the suggestion that taxonomic categories more isolated on the phylogenetic scale be given higher priority can be considered little more than a gloss on 'diversity.' It is a decision that genetic diversity is more important than diversity defined in terms of the sheer number of species… Even here, however, the pervasive problem of ranking values begins to obtrude. Reasonable individuals could differ in their value judgments when faced with the choice, for example, between saving several species that have near neighbors and one species that has none, if the costs are the same in each instance. This choice amounts to ranking the value of species diversity against the value of genetic variability. But, while reasonable individuals could differ, it is not unreasonable to hope that a goal can be agreed upon: the problem, ultimately, is only one of choosing a single concept of diversity to pursue" (Norton 1987, 255–256).

populations seems to betray rather than manifest the intended content of the concept.[4] Something narrower is clearly the target. But what, exactly? As the preceding section indicates, biological sciences do not seem to pinpoint a more specific content for the concept and offer little guidance about which diversities should have priority in doing so. Characterizing the concept of biodiversity necessitates some assessment of what is relevant on pain of vacuity. But from where, what information source and/or authoritative standard, will those relevance assessments come?

Norton plausibly identifies a pervasive but thoroughly misleading assumption as having hindered the search for an adequate definition, in this case and much more widely, especially in philosophically loaded contexts. The insidious assumption in question is that "our categories 'correspond' to an existing, 'deeper' reality" (Norton 2006, 12). From mainstays of the history of philosophy – Platonic forms, Aristotelian essences, Kantian categories – to the contemporary faith that a rational and yet somehow scientifically inscrutable faculty of intuition tracks deep truths about reality (Machery 2017), the idea that human thought and language must mirror mind-independent contours of objective reality maintains a firm grip on the philosophical community, at least within the so-called analytic tradition. But in this case, "there is no 'correct' biological definition to be 'found,' as one might discover a gem under a rock" (Norton 2006, 13). For Norton, pragmatism provides the salutary antidote: "communicative usefulness, not truth-by-correspondence, should determine our definitions. But usefulness implies we must carefully examine the shared purposes of the communication – and that leads back to the subject of social values and commitments ... we are looking for a definition that is *useful* in deliberative discourse on how to preserve biological diversity, however defined" (13).[5] Thus, for a concept such as biodiversity, "we are forced to conclude that we cannot know what we mean until we know what we care about" (12).

[4] On the other hand, the idea that "biodiversity" should include the genetic diversity required to ensure species persistence, however that may be characterized, seems compelling. In general, adequately characterizing biodiversity involves considerations of not only how it is synchronically represented, but also its diachronic representation, that is, its persistence. This norm is especially compelling on some value-driven approaches to defining the concept (see below).

[5] Here it is worth highlighting the close affinity with Rudolf Carnap's thoroughly pragmatic method for determining concepts called explication (Carnap 1962, chapter 1; Justus 2012b; Brun 2016).

This eminently reasonable proposal, however, introduces a few flies into the ointment. Grappling with a simple but crucial question reveals the trouble: How is usefulness gauged? As Norton notes above, usefulness ultimately depends on values and commitments, in this case those relevant to saving biodiversity. But then at least two serious difficulties arise. The first is methodological. Usefulness to deliberation about preserving biodiversity is supposed to determine how "biodiversity" is defined. This suggestion, however, seems blatantly circular or vacuous: the content of the concept already features in the procedure being employed to define the concept. Without some inkling about what biodiversity is first, how can one gauge what is deliberatively useful in preserving it?

Norton seems to recognize this potential problem. He clarifies that it is what humans value about the biological world, rather than some pre-theoretic notion of biodiversity itself, that features in the procedure for assessing the relevant usefulness. This helps circumvent the apparent circularity but on pain of raising another, equally problematic methodological worry. With values playing this role, what is being defined looks to be a different concept, something like "what we value about the biological world," rather than the original target, biodiversity. In the interests of full transparency, the proposal would then just replace biodiversity with another concept, rather than attempt to define it.

At times, Norton seems to betray precisely this intention and bite the bullet. Biodiversity, he states, "has come to function as a label for the broad concerns for nature, its life forms, and its processes" (2006, 11). If "biodiversity" does indeed function as such a label, the relevance of values is manifest. If there is confusion or hesitation about this relevance, Norton might counsel, it is likely due to a residual realism about concepts, to a thoroughly non-pragmatic and misguided view that there is some mind-independent biodiversity concept to be discovered divorced from how humans value the natural world. Perhaps this is the best that can be expected for ethically charged concepts for which the biological sciences, perhaps sciences generally, do not pinpoint a determinate content. Not all concepts "carve nature at the joints," and something must take up the conceptual slack. But one need not be a table-pounding realist to worry about the role of values here. Values are, in general and especially concerning the natural world, complicated, heterogeneous, and sometimes outright incompatible.

The potential problem this complexity and conflict might pose is well known to Norton (see Norton 1994). The two main opposing environmentalist positions – anthropocentrism and the view that the natural world is intrinsically valuable, that is, valuable independent of any value it may have to humans or other valuing creatures – "characterize values in incommensurable ways," which leads to "discourse [being] characterized by disagreements about the nature of the problem, the goals to be pursued, and the nature of the social values considered worth protecting" (2006, 13). To bypass this "turf war," Norton advocates a value pluralism that acknowledges that humans value the natural world in many different ways, all of which he says are nevertheless "human values" (14). That pluralism, he believes, is reflected in the motley set of values recognized in the Endangered Species Act (1973): aesthetic, ecological, educational, historical, recreational, and scientific. Once this diverse set of values is adopted, Norton believes perennial, philosophically charged debates about value would cease to plague biodiversity conservation efforts (or a least impede them less severely). The focus would (hopefully) shift toward more pragmatic considerations of more practical goals and how they can best be achieved.

Some of Norton's own commitments cut against this pragmatic optimism. For instance, Norton emphasizes that none of the values cited in the Endangered Species Act can be plausibly interpreted as attributing intrinsic value. But this absence then poses a serious threat to the value pluralism he wants to endorse. From the perspective of intrinsic value advocates, the ESA values would appear to bypass rather than pluralistically respect their commitments. Norton also rightly notes that "intrinsic value claims are largely unresolvable by experience or empirical evidence" (2006, 13–14) and are therefore ideological stumbling blocks that typically derail policy deliberations.[6] The unresolvability by experience or empirical evidence then seems to show that such claims cannot be pluralistically integrated into these policy debates. As Norton indicated earlier, but the invocation of pluralism apparently disregards, disparate values can and often do produce very different views about what the problems are and what the goals should be when attempting to conserve biodiversity. Rather than remove obstacles to determining the biodiversity concept, the focus on values, even when dressed in pluralistic language, seems to introduce more obstinate ones.

[6] See Justus et al. (2009a, 2009b), and Sagoff (2009) for a detailed discussion of this issue.

## 3 Implicitly Defining Biodiversity

If ethical values play an important, if not indispensable, role in adequately characterizing biodiversity, the key question is how the difficulties of value disunity and intractability Norton's approach encountered can be avoided. One ambitious option would be to wade directly into the thorny ethical issues about the value of the natural world and attempt to show there is a uniquely correct account. The perennial and, for someone like Bernard Williams, quintessential absence of intersubjective agreement in ethics (Williams 1986) – a discord that seems *more*, not less pronounced with respect to the value of the natural world – casts a pessimistic pall on that project. What is needed is some way to regiment the role values play, to marshal or perhaps just circumvent the disunity, discord, and apparent intractability characteristic of debates about the value of the natural world into a source of manageable content about the concept of biodiversity. Fortunately, there is a scientifically oriented alternative endeavoring to achieve this winnowing of content. Although ecology and other long-established biological sciences appear to offer little guidance (see Section 1 above), attending to the ethically driven science most directly concerned with saving biodiversity, conservation biology, furnishes a novel definitional possibility: an implicit definition.

The strategy, first pursued by Sarkar (2002, 2005), lets scientific practice rather than direct ethical theorizing delimit the role values play in characterizing biodiversity. Specifically, values serve as inputs for algorithmic methods used by conservation biologists to prioritize places to save biodiversity. These place-prioritization algorithms are then taken to "implicitly define" what biodiversity is. Before the strengths and weaknesses of this strategy can be assessed, some background is needed.

About the time the term "biodiversity" came into vogue in the mid-1980s, conservation biology emerged as a scientific discipline distinct from ecology devoted principally to the conservation of biodiversity. A widespread concern that anthropogenic actions, especially deforestation in the tropics, was producing a significant increase in the species extinction rate precipitated its emergence (Soulé 1985). From the very beginning, the primary (but not exclusive) goal of the new discipline was the design of conservation area networks, such as national parks, nature reserves, and managed use zones, to protect biodiversity from anthropogenic transformation. But different ethical views about which features of the natural world are valuable generated contrasting views about what should be the target of conservation efforts, which in turn generated distinct methodologies within conservation biology.

Early on, two general approaches to conservation loosely characterized the discipline, labeled the "small-population" and "declining-population" paradigms in an influential review (Caughley 1994). The powerful legal framework of the Endangered Species Act and National Forest Management Act (1976) helped catalyze the former, which originated and was widely adopted in the United States. Extinction threats to individual species were its main focus. By analyzing their specific ecologies, including habitat requirements and geographical distributions, many conservation biologists thought "minimum viable populations" could be found: population sizes below which stochastic processes would significantly increase the probability of extinction. It quickly became clear, however, that this aim had insurmountable methodological obstacles. Much more data than feasible to collect for most species was needed (Caughley 1994), and the interspecies dynamics essential to most species' survival were largely ignored (Boyce 1992).

The declining-population paradigm that came to dominate Australian conservation biology focused on deterministic, rather than stochastic, causes of decline. Unlike the small-population paradigm, the goal was discovering and eradicating these causes before stochastic effects became significant. Since one of the drivers – if not the main driver – of population decline was (and is) habitat loss, a principal concern was preserving a region's full complement of species diversity (and required habitat) in protected areas, rather saving specific species. With meager monetary resources at their disposal, conservation biologists rightly concentrated on developing methods that maximized representation of that diversity while minimizing the total area (and hence cost) required to do so.

The methods that achieve those goals are called place-prioritization algorithms (Margules and Sarkar 2007) or, sometimes, reserve-selection algorithms, somewhat misleadingly because they are often employed to identify non-reserve status protected areas. Place prioritization involves solving a resource allocation problem; priority must be given to protecting some areas over others given limited budgets. Doing so requires geographically referenced data for the region of conservation interest. Specifically, a region is "grided" into a set of individual places called cells ($c_j, j = 1, \ldots, n$) that can be different areal sizes ($a_j, j = 1, \ldots, n$). A set of biodiversity surrogates **S** ($s_i, i = 1, \ldots, m$)

are identified for the region (more below) with representation goals or targets, one for each surrogate ($t_i$, $i = 1, \ldots, m$). For example, ensuring each surrogate is represented in at least one cell is a typical target. Finally, there is a set of probabilities of finding $s_i$ at $c_j$ ($p_{ij}$, $i = 1, \ldots, m; j = 1, \ldots, n$). If presence/absence records for the surrogates are available, these probabilities are extremal (1 for presence, 0 for absence). Examples of such records include distributional data sets of plants and animal species derived from surveys or even historical records. Nonextremal probabilities are often generated from species distribution models that probabilistically predict species distributions in a given region. The allocation problems for which this information is relevant can take different forms. Two common problems are (1) find the set of cells with the smallest area such that every representation target is satisfied and (2) find **M** cells that maximize the number of representation targets satisfied (usually because only **M** cells can be protected).[7]

These problems can be analyzed and solved algorithmically, in a variety of ways (Sarkar et al. 2006). There are so-called exact algorithms that find optimal solutions to problems like (1) and (2), assuming one exists. So-called heuristic algorithms, labeled "heuristic" because they do not guarantee optimal solutions, prioritize areas for protection by iteratively applying a hierarchical set of conservation criteria such as rarity, richness, or complementarity. Rarity, for example, requires selecting the cell containing the region's most infrequently represented surrogates, thereby helping ensure any endemic taxa are represented. Given the importance of protecting endemics, rarity is therefore often used to select the first set of cells in a place prioritization. Richness, on the other hand, requires selecting the cell containing the most surrogates. If one suspects that selecting the most first might ultimately yield maximal surrogate representation, or sufficient representation in minimal area, richness is often used. More than any other

[7] These are the so-called expected surrogate set covering problem (ESSCP) and maximal expected surrogate covering problem (MESCP), respectively (Sarkar et al. 2006). They can be stated exactly. If two indicator variables $X_j$ ($j = 1, \ldots, n$) and $Y_i$ ($i = 1, \ldots, m$) are defined:

$$X_j = \begin{cases} 1, \text{if } c_j \in C; \\ 0, \text{otherwise}; \end{cases} \quad Y_i = \begin{cases} 1, \text{if } \sum_{c_j \in C} p_{ij} > t_i; \\ 0, \text{otherwise}, \end{cases}$$

where $C$ is a set of candidate protected areas, then ESSCP is the problem: minimize $\sum_{j=1}^{n} a_j X_j$ such that $\sum_{j=1}^{n} X_j p_{ij} \geq t_i$ for $\forall s_i$. MECSP is the problem: maximize $\sum_{i=1}^{n} Y_i$ such that $\sum_{j=1}^{n} X_j = \mathbf{M}$; where **M** is the number of protectable cells.

criterion, however, the efficacy of heuristic algorithms is due primarily to complementarity, which requires selecting subsequent cells that complement those already selected by adding the most surrogates not yet represented (Justus and Sarkar 2002). Place-prioritization algorithms demonstrate how limited funds can be efficiently marshaled for biodiversity conservation, and in so doing they have been quite effective in securing actual funds in policy-making contexts (Kingsland 2002a).

There is, of course, a cost to employing these algorithms. Unlike general theories such as island biogeography theory (MacArthur and Wilson 1967) that abstract from the particularities of individual areas, place prioritization algorithms require geographically explicit data on the surrogates taken to represent biodiversity. Acquiring such data directly via field surveys is usually impossible given limited time, money, and relevant expertise with which to complete them. This difficulty has spawned an extensive research literature attempting to find a set of biotic or abiotic attributes (e.g., specific species, soil types, geological features, vegetation types) whose geographical distributions are at least feasibly and ideally cheaply acquirable, for instance, via remote sensing lidar or easy-to-execute species-specific surveys (Margules and Sarkar 2007), that would adequately represent biodiversity. Sarkar (2002) labels these attributes "estimator surrogates." The "adequate representation" in question can be gauged by how well place-prioritizations using them accord with place-prioritizations using "true surrogates" taken to represent biodiversity in general (see Sarkar et al. 2005). Species diversity or species at risk are common true surrogates in actual conservation-planning contexts, but given the supreme onerousness of acquiring distributional data for *all* species or even species at risk in a region, proper subsets are almost always used instead.

Place-prioritization algorithms provide a powerful tool for *conserving* biodiversity, but how is this relevant to the task of *defining* biodiversity? The connection is made, Sarkar (2002, 2005) argues, with the concept of implicit definition.

Explicit definitions are philosophically familiar, not least as the aspirational output of traditional conceptual analysis. By eliciting and pumping intuitions about hypothetical scenarios, usually quite fantastical ones – Did alien-manipulated Alan freely act to push the button? Does swamp man believe X? Is water XYZ on twin earth? – conceptual analysis was billed as the method for uncovering necessary and sufficient conditions for many (most? all?) concepts, conditions that could then be encapsulated in explicit definitions. "Knowledge is justified true belief" and "Bachelors are unmarried males" are the most well-worn examples in philosophy, but there are many others. The conceptual nuance and argumentative sophistication exhibited by practitioners of traditional conceptual analysis is matched only by its dismal track record. Almost no concepts, and certainly no philosophically substantive concepts, yielded to the methodology with successful explicit definitions.

But there is an alternative to explicit definition, one that can avoid these deficiencies and offer a more promising method for characterizing some concepts: implicit definition. Some domains of inquiry seem to be exhaustively and definitively captured by a (finite) set of principles or statements. Commonly cited examples include logical and mathematical domains, such as the calculus of first-order logic, the axioms of Peano arithmetic, Euclidean geometry, and set theory, as well as empirical domains, such as axiomatic approaches to quantum mechanics or relativity theory. The idea, then, is that these principles or statements themselves implicitly define the expressions they contain. For instance, the meanings of "line" and "point" are determined by the axioms of Euclidean geometry, and the rules of introduction and elimination in first-order logic implicitly define the relevant logical constants (see Hacking 1979). Depending on one's favored theory of meaning (and hence account of implicit definition), on standard truth-theoretic accounts the expressions would mean whatever is necessary for those principles or statements *to be true*; on use-theoretic accounts they would mean whatever *regarding* those principle or statements *as true* requires (see Horwich 1998, chapter 6).

Implicit definitions raise complex and deep philosophical issues: whether and when implicit definitions might be judged to yield a priori knowledge, when they might be considered analytic, and when they satisfy what is now considered the classical theory of definition, its conservativeness and eliminability conditions in particular, and when failing to satisfy that theory is justifiable. Only implicit definitions in formal domains have any real prospect of possessing these properties, which is why discussions of implicit definition are much rarer in empirical domains. Proposed implicit definitions of empirical concepts, for example, almost certainly violate the conservativeness and eliminability conditions (see Gupta 2015). Put informally, that means that these candidate implicit definitions are not simply innocuous instances of Byzantine paraphrase. When added to a theory or language, they can do real work, such as facilitating new predictions or deductions of new theorems (i.e., they can be *nonconservative*), and they can alter probability attributions, change deducibility relations, and in general semantically change and enrich what can be expressed (i.e., they can be *ineliminable*). The implicit definitions mentioned above, and almost certainly *all* philosophically substantive implicit definitions, are nonconservative or ineliminable or both. But if they fall outside the established grounds of the traditional theory, what exactly is their justificatory basis?

This is a critical question at the very heart of philosophical methodology, and addressing it fully is obviously orthogonal to our eco-philosophical aims (see Horwich 1998; Hale and Wright 2001; MacBride 2003). But its importance reverberates against the proposed implicit definition of biodiversity. In several significant ways the proposal diverges from the canonical kinds of cases cited above, which makes the issue of justification all the more pressing.

Consider the multifarious nature of these algorithms. Unlike Euclidean geometry or Peano arithmetic, there is no single stable set of principles or statements facilitating the proposed implicit definition of biodiversity. Rather, algorithms are supposed to be doing the defining, and they are numerous. Selection criteria are one source of algorithmic diversity; places can be prioritized by many distinct selection criteria. They include rarity, richness, and complementary as described above, but there are several others: adjacency (which favors selecting cells adjacent to those already selected), area, compactness (which favors clustering by selecting cells closest by some distance metric to those already selected), cost (which favors cells cheaper to protect when such economic information is available), ecosystem importance (which considers the role the constituents of cells may play in broader ecosystem dynamics, such as ecosystem services), social value (which can ensure preference for cells with aesthetic, cultural, historical, or even recreational appeal), vulnerability (which considers how likely it is cells will be transformed in the future, for example, by development), and others. This multiplicity raises a family of related concerns.

Across different contexts distinct algorithmic configurations of these criteria are typical, and they almost always yield different prioritizations. Depending on the configurations, they may be *very* different prioritizations. Algorithms that initialize by rarity and then iteratively maximize complementarity will likely select very different sets of cells than ones that include compactness, ecosystem importance, or social value as additional criteria.[8] Even the same criteria can and usually do yield different prioritizations when hierarchically ordered in different ways with heuristic algorithms. And the prioritizations will be very different with different targets or surrogates.

Perhaps recognizing the challenge this multiplicity poses, Sarkar (2002, 148) suggests, "Different concepts of biodiversity are implicitly defined by these algorithms: biodiversity is the relation used to prioritize places." Before worrying about the character of the family of concepts mentioned in the first clause, problems with the second claim (and its tension with the first) should be addressed. First, if each algorithm implicitly defines a different biodiversity concept, and there are many distinct algorithms, and especially if those algorithms almost always produce different prioritizations, in what sense could any singular relation, let alone *the* biodiversity relation, be doing the prioritizing? Note that for exactly the same set of places (the area of conservation interest), the same true and estimator surrogates, the same conservation targets, and the same distributional data, different algorithms with different selection criteria (or the same criteria hierarchically ordered differently) will nevertheless almost always prioritize differently. And of course the surrogates and targets usually vary with the goals the algorithms are utilized to accomplish; this is yet another source of variation in how places are prioritized. The sheer profusion of different algorithms and prioritizations they generate casts considerable doubt on the idea that some singular biodiversity relation underlies prioritization.

Rather, what is clearly responsible for different prioritizations are the specific algorithmic configurations (selection criteria, surrogate choices, targets, etc.). And as the heterogeneous list of selection criteria described above shows, many facets of those configurations have little to do with biology and, hence, it seems, with *bio*diversity. Adjacency, cost, and cultural value are crucial considerations when determining how to best allocate limited conservation funds, but this import is socioeconomic, not biological. If "biodiversity is to be (implicitly) defined as what is being conserved by the practice of conservation biology" (Sarkar 2002, 132), that practice definitionally favors cost-effective conserving as well as other things that humans value that have little to do with the biological world.

[8] Even seemingly quite similar algorithms, such as rarity-complementary and richness-complementary algorithms, usually select disparate cells (Margules and Sarkar 2007).

If practice is the guide, the "bio" in "biodiversity" is as inapt as the "biology" in "conservation biology."

Absent a convincing case that "biodiversity is the relation used to prioritize places," the notion that "each place-prioritization algorithm implicitly defines a (slightly) different concept of biodiversity" (Sarkar 2002, 135) loses traction. Given the variety of algorithms employed, what justifies the belief they implicitly define concepts *of biodiversity*? Some standard for adjudicating the claimed conceptual linkage is needed. That standard cannot derive from any pre-theoretic understanding of biodiversity given its vacuous generality, hence the turn toward implicit definition. The only answer that seems available, that really unifies the disparate algorithmic applications across conservation contexts, is human valuation. What is being prioritized is what we value about the "natural" world. This certainly tracks various features of the biological world, but also obviously selectively. And commingled throughout these assessments are assorted socioeconomic considerations, hence the patently nonbiological character of so many selection criteria. The problem of changing the goalposts that confronted Norton's approach to defining biodiversity has been recast in different terms, but not resolved.

There is perhaps an even more pressing prior concern: Why think anything at all is being (implicitly) defined in place-prioritization? Underlying canonical cases of implicit definition is a presumption that the statements or principles doing the defining are true, or have whatever authoritative surrogate of truth one's epistemology or metaphysics sanctions. What then confers meaning on terms is the meanings of those (true or true-like) statements. Since they authoritatively (and typically exhaustively) "capture" whatever content there is in these domains (arithmetic, classical logic, Euclidean geometry, set theory), the contained terms must mean whatever is required to capture that content. But note the glaring disanalogies with conservation biology. First, prioritization algorithms are nowhere near exhaustively spanning all the methods and practices employed by conservation biologists to protect biodiversity. Ecological restoration, ecosystem and community modeling, endangered species rehabilitation procedures, population viability analyses, survey and sampling techniques, threat detection, and targeted intervention are but a proper subset of what conservation biologists do. No one doubts that place-prioritization is an indispensable part of their practice, but why should only this portion be definitionally potent, assuming any is? (See below.) The narrow focus on algorithms looks like cherry-picking.

There is a second sense in which the focus seems too narrow. On the implicit definition approach, many other entire disciplines plausibly have definitional potential for biodiversity. Recall that at its most inclusive, biodiversity's scope is the biological world in all its intricate and fascinating manifestations. There are obviously several sciences devoted to understanding this world: biogeography, developmental biology, ecology, evolution, molecular biology, and so on. It is thoroughly unclear why the information and insights these sciences uncover are irrelevant to the task of defining biodiversity. This is not to deny that the problem of vacuous generality has to be avoided; it does. The question is why place-prioritization algorithms in particular, and conservation biology more generally, is the (unique?) way to do so.[9] Each of these other sciences (and their numerous subdisciplines) also concerns only part of the biological world. They each therefore offer a potential solution to the vacuity problem. The ultimate goal of conservation biology is conserving biodiversity; the ultimate goal of these sciences is understanding biodiversity. What makes the goal of the former definitionally privileged over the latter remains elusive.

A third disanalogy might be the most problematic. In canonical cases of implicit definition, the authoritative meanings of statements or principles confer meaning on the terms they contain. The meanings in question are descriptive and truth-functional on standard semantic accounts. But the "meaning" of a prioritization algorithm, assuming such a peculiar application of that label is intelligible, is not straightforwardly descriptive but akin to something prescriptive (or conditionalized prescriptive). Effectively, each algorithm is a specific if-then procedure. With specific representation goals (targets), surrogate choices, geographical information (distributional data), and hierarchically structured selection criteria, an algorithm reveals which areas should be protected to achieve the targets subject to the constraints. How these apparent instances of potential instrumental rationality confer meaning on "biodiversity" is entirely unclear. It is not just that the term "biodiversity" occurs nowhere in these complicated if-then conditionals –

[9] "[B]iodiversity *must* be analyzed in the context of conservation biology and what it, as a goal-oriented enterprise that prescribes policies, must accomplish as it tries to conserve biodiversity" (Sarkar 2002, 132; emphasis added). The justificatory force behind the "must" is the issue.

unlike the correlate case in implicit definitions for "line," "successor," "set," and so on – although that does pose a serious obstacle. Without access to a nonvacuous pre-theoretic concept of biodiversity that would establish the connection – which, of course, is why implicit definition is being pursued in the first place – there is a conspicuous gap between what place-prioritization algorithms deliver and the notion of biodiversity.[10] The problem also does not seem to be that standard theories of semantics are too confining, and that a new, more pliable theory would do the trick. Rather, the algorithms simply seem to furnish the wrong kind of (prescriptive) content to define biodiversity (assuming they furnish any such content), a concept that is supposed to describe some proper part of the biological world. It's as if one were expecting engine enhancement instructions to define what an automotive "hot rod" is. The proposed conceptual bridge seems a bridge too far.

These are differences that make a difference. Divergences from well-known implicit definitions in the case of biodiversity reflect more than differences of context or incidental detail. They reflect departures from the very conditions that render proposed implicit definitions defensible. Implicit definitions of biodiversity unfortunately appear to fare as poorly as explicit ones.

## 4 Conclusion

In a recent incisive series of papers, Carlos Santana (2014, 2016, forthcoming) comes to a similarly negative appraisal of biodiversity on eliminativist grounds. Specifically, biodiversity realism is indefensible. The relevant realism can be cashed out in at least three ways: as a homeostatic property cluster, as a multiply-realizable functional kind, or as an indispensable explanans and explanandum in different fields of biology. As you might expect from the preceding section, the three accounts fail in unique ways, but the common poison pill is the motley, disunified assortment of things that are supposed to comprise biodiversity:

> Multidimensional biodiversity includes not only the number of different species (species richness) and their relative abundance, but also genetic diversity, phenotypic diversity, differences in evolutionary history such as

[10] The facile answer that prioritization algorithms are intended to help protect biodiversity falls flat for the reasons discussed earlier.

> phylogenetic diversity, diversity at community and ecosystem levels (e.g. ecosystem diversity, functional diversity, and trophic network diversity), and differences in the diversity represented between areas (complementarity and β-diversity), among other things. (Santana forthcoming, 15)

Such a Frankenstein conglomeration will be unpalatable for those inclined toward scientific realism, particularly if explanatory utility is tightly coupled to the quality of being a scientific kind and the carving of nature at its ontologically supple joints (see Santana forthcoming, n. 1). Perhaps a more pragmatic take on explanatory value and other functionality would make room for such an assorted scientific creature. That remains *possible.* Then again, it looks quite implausible that a pragmatist will ever uncover a dimension of practical utility the realist missed given the glaring deficiencies already exposed. Despite our different backgrounds, sometimes we have to come together and recognize that a creature is a monster, and eliminate it.

# 6 Progress in Applied Ecology

Ecology has received far less attention from philosophers of science than other areas of the life sciences, such as developmental, evolutionary, and molecular biology and systematics. But that tide is turning, not least because societal concerns are forcing their hand. The severity and complexity of environmental problems, and ecology's potential role in helping solving them, has become much more apparent as of late. Global threats to coral reefs and the ecologically informed management strategies developed in response are vivid examples. This chapter explores the range of ways values influence both applied ecological science and policy making informed by that science, and the intricate but often underappreciated philosophical issues they raise.[1]

We begin with an unmitigated success story of scientific progress. As a distinct discipline of applied ecology, conservation biology emerged as a rigorous science focused on protecting biodiversity in the 1980s. Two algorithmic breakthroughs in information processing made this possible: place-prioritization algorithms (PPAs) and geographical information systems (GIS). They provided a defensible, data-driven methodology for designing reserves to conserve biodiversity. This obviated the need for largely intuitive and highly problematic appeals to ecological theory to design reserves at the time. They also supplied quantitative, critical assessments of existing reserves. Most reserves had been designated on unsystematic, ad hoc grounds and consequently poorly represented biodiversity. Demonstrating this convincingly was unsurprisingly crucial to ensuring biodiversity would be adequately protected in future policy-making contexts.

[1] For a complementary analysis of these issues in the agricultural sciences, see Thompson (2011). Similarly, Oppenheimer et al. (2019) incisively explore the philosophically rich relations between science and policy for several large-scale environmental problems.

Despite these unquestionable advances, that they constitute scientific "progress" has been criticized. Ecological theory, it is claimed, is required for genuine progress about reserve design; algorithmic innovation in data processing is insufficient (see Linquist 2008). Place-prioritization algorithms (PPAs) are also supposedly less scientifically grounded and produce reserves that poorly protect biodiversity. This chapter argues that on all accounts this criticism is indefensible. It involves numerous inaccuracies about the science, misconstrues the character of applied science, and, most crucially, relies on an untenable conception of progress for applied sciences with ethical objectives such as conservation biology. Although applied sciences are unquestionably science and employ scientific methods, what constitutes progress within them should not always, and definitely not in this case, be judged by the standards thought appropriate for classic descriptive sciences such as chemistry, evolutionary biology, and physics. Existing philosophical accounts of scientific progress that exclusively emphasize acquiring knowledge or ontologically uncovering the "true nature" of reality (e.g., Bird 2007) must be rethought and broadened to recognize the ethically driven character of some applied sciences.

## 1 A New Discipline of Applied Ecology Emerges: Conservation Biology

The newly energized environmental movement of the 1960s thrust ecology into the social limelight (Nelkin 1977).[2] The 1970 and 1971 presidential addresses of the Ecological Society of America reflected the sea change:

> Ecology has been pulled out of the shadows and thrust upon the central stage. (Bormann 1971, 4)
>
> In the last three years [the discipline of ecology] has achieved a degree of fame (or notoriety) far exceeding our most extravagant hopes and dreams of a decade or more ago. (Auerbach 1972, 205)

This exerted considerable pressure on ecologists to remedy the environmental problems being revealed to the public. Among them, species loss due

[2] It is hard to overstate the cultural significance and societal power of the environmental movement. The legislation the movement catalyzed, to highlight just one kind of example, was unprecedented. In the United States alone, it included the Clean Air Act (1963), Wilderness Act (1964), National Environmental Policy Act (1970), Endangered Species Act (1973), Forest and Rangeland Renewable Resources Planning Act (1974), and Clean Water Act (1977).

to tropical rainforest deforestation was being heralded as a crisis in numerous high-profile scientific publications (e.g., Gómez-Pompa et al. 1972; Ehrenfield 1976; Ehrlich and Ehrlich 1981; Soulé 1985; Janzen 1986). A strategic institutional response was called for, and a new discipline emerged: conservation biology.

## 1.1 *Early Reserve Design: Theory-Heavy and Data-Light*

Given the magnitude of deforestation, ecologists felt uniquely obliged and qualified to evaluate the threat and to identify reserve areas to protect portions of the diversity of biological entities and phenomena, or "biodiversity," the rainforests contained (see Chapter 5).[3] Ideally, the reserves would maximize representation *and* persistence of the biodiversity they contain. The challenge was developing a method to identify such reserves. This was especially daunting since data on distributions of most species did not exist. At the time, a handful of detailed distributional data sets had been developed. Myers (1988), for example, estimated that only 0.5 million of a predicted 2.5–30 million species in tropical forests alone had even been identified; pinpointing areas of high tropical species richness for protection, therefore, seemed doomed to fail. And without adequate data, selecting areas for protection also seemed doomed to indefensible ignorance about what they contained.

Furthermore, little was known about the autecology of the vast majority of species – the environmental factors affecting plants or the habitat requirements of animals – especially for tropical species. Beyond a scattering of species with economic value, or those garnering the attention of professional biologists as laboratory model organisms or targets of fieldwork, the only possible exceptions were some species of temperate birds, butterflies, and large mammals, owing primarily to their charismatic appeal. Ecologists also knew little about species interactions, specifically, dependency relationships between them. Hence, they could usually only speculate that protection of some species would ensure adequate protection of another species. In the mid-1970s, the overall lack of ecological information relevant to conservation planning prompted David Ehrenfeld (1976, 652), one of the

[3] Ecologists were obviously also protecting their livelihood. As one tropical ecologist provocatively implored, "If biologists want a tropics in which to biologize, they are going to have to buy it with care, energy, effort, strategy, tactics, time, and cash" (Janzen 1986, 306).

founders of the US Society for Conservation Biology, to remark that "the population dynamics and management ecology of nearly all species are still largely unknown."

There was, however, a brand new and impressively mathematical ecological theory that seemed to exactly fit the methodological bill: the equilibrium theory of island biogeography (MacArthur and Wilson 1963, 1967). Several prominent American ecologists responded to the dearth of data difficulty by using this theory to champion general principles of reserve design (Terborgh 1974; Willis 1974; Diamond 1975; Wilson and Willis 1975; Diamond and May 1976). They assumed, to make the theory applicable, that for most species protected areas surrounded by degraded habitat – for example, patches of rainforest protected from clear-cutting – were ecologically similar to islands. According to island biogeography theory, these "ecological islands" would hold more species the larger and closer together they were. Specifically, the theory predicts that species richness is determined by an equilibrium between extinction and immigration processes; increases in island isolation decreasing immigration or decreases in island area increasing the extinction rate will produce equilibria with fewer species. That island biogeography theory may provide "geometrical rules of design of natural reserves" (Wilson and Willis 1975, 528) was first proposed by Edwin Willis as early as 1971 (Willis 1984) and independently by Edward Wilson around the same time (Wilson 1992).

Without a doubt, the most influential analysis to use of the theory in reserve design was Diamond (1975). Adapting a diagram from Wilson and Willis (1975), and choosing a somewhat weaker label of "principles" rather than "rules," Diamond (1975, 143) proposed six design principles (Figure 7), "derived from island biogeographic studies."

Diamond (1975) justified each principle by appealing to two factors: minimization of population extinction rates of populations in reserves, or maximization of immigration rates. For instance, the closer proximity of protected areas with the same habitat required by principle C arguably helps ensure greater immigration rates between populations they contain. This, in turn, arguably helps ensure individuals will emigrate to areas with declining or extinct populations.

But despite its theoretical pedigree, only principle A proved uncontroversial. Principles B–F were highly contentious. Principle B, for example, initiated the vituperative controversy over whether, *in general*, a single large or

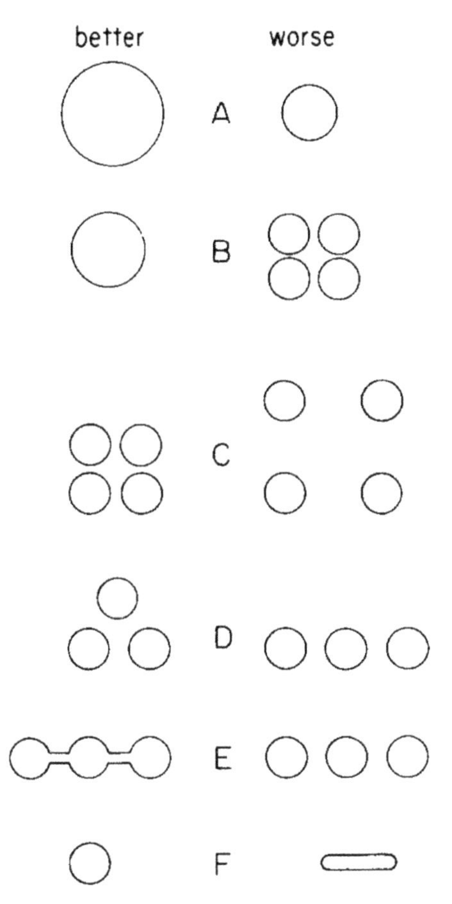


Figure 7 Diamond's proposed geometric design principles for conservation reserves. Principles B, D, E, and F are from Wilson and Willis (1975). Adapted from Diamond (1975, 143)

several small ("SLOSS") areas of equal total area would protect more species, labeled the SLOSS debate by Simberloff and Abele (1982). Diamond (1975, 144), however, qualified only B by noting that:

> separate reserves in an inhomogeneous region may favour the survival of a different groups of species; and that even in a homogeneous region, separate reserves may save more species of a set of vicariant similar species, one of which would ultimately exclude the others from a single reserve.[4]

[4] Diamond (1975) did not clarify the meaning of "homogeneous" or consider problems involved in classifying different habits. And contra Diamond's (1975) assumptions, most regions are not habitat homogeneous (Margules et al. 1982).

Similarly, principle E initiated the contentious debate over the conservation value of habitat corridors (Simberloff et al. 1992). Although Diamond (1975, 144) only claimed habitat corridors "*may* significantly improve the conservation function of [reserves]" (emphasis added), there was little empirical support for this claim or appreciation of how difficult its acquisition would be (Nicholls and Margules 1991).

Simberloff and Abele (1976), two prominent members of what was later memorialized as the "Tallahassee Mafia" (see Dritschilo 2008), were the earliest and most visible critics of Wilson and Willis's (1975) "rules" and Diamond's (1975) "principles." They argued, contra principle B, that "the" species–area curve:

$$S = kA^z,$$

where $S$ is the number of species in area $A$, and $k$ and $z$ are constants – which played a substantial role in the development of island biogeography theory through the work of Frank W. Preston (1962) – did not unequivocally support single large areas. Specifically, whether a large area or several small ones contain more species at equilibrium depends on (1) the proportion of species the latter share and (2) the exact shape of the species–area curve. Existing data, they claimed, showed the shape varies markedly with taxon. Accounts of the curve treating all taxa as equal, such as by island biogeography theory, were indefensible. Depending on the taxa in the region, therefore, a large area may contain much fewer species than several small areas. In support, Simberloff and Abele (1976) presented data from a red mangrove (*Rhizophora mangle*) and suggested the archipelago was not atypical in contradicting principle B. They also pointed out, rather astoundingly, that island biogeography theory itself implies species gradients in dispersal and survival can *favor* several small over a single large area (Simberloff and Abele 1976).

Subsequent studies reinforced these criticisms and added others (see Pickett and Thompson 1978; Abele and Connor 1979; Kushlan 1979; Gilbert 1980; Higgs and Margules 1980; Higgs and Usher 1980; Higgs 1981; Margules et al. 1982; Simberloff and Abele 1982). For example, whether nondegraded habitat "islands" are even ecologically similar enough to oceanic islands to make island biogeography theory applicable depends on the species and degree of degradation (Margules et al. 1982). Often the dissimilarities were glaring and significant, and too often judging similarity was entirely unclear. By the mid-1980s, it had become clear that the initial promise of island biogeography theory in reserve design had been seriously overestimated. In an exhaustive review, for instance, Gilbert (1980) argued there was insufficient empirical support for island biography theory itself. Thus, Margules et al. (1982) stressed, design principles supposedly based on it were unjustified. The "scientific revolution" (Diamond 1975, 131) catalyzed by MacArthur and Wilson (1963, 1967) proved unhelpful when, "For a variety of taxa, for a number of different habitat types, and for a wide range of sizes of biota as a fraction pool, either there is no clear best [reserve design] strategy, or several small sites are better than one large site" (Simberloff and Abele 1982, 48).

Disagreement about the goal of reserve design also complicated the issue. The goal of Diamond (1975) and Wilson and Willis's (1975) design criteria was protection of maximum species richness when a reserve reached equilibrium after its ambient area was degraded, not protection of the particular species it currently contains. An implication of island biogeography theory focused attention on the former: "although the *number* of species on an island may remain near an equilibrium value, the *identities* of the species need not remain constant, because new species are continually immigrating and other species are going extinct" (Diamond 1975, 134–135). From this perspective, the specific species composition of an area is irrelevant to reserve design since it would change over time, especially as a reserve reaches a new equilibrium after its ambient area is degraded. Yet, attempting to maximize species richness, Margules et al. (1982) criticized, may leave many species unprotected now, or fail to protect them efficiently in the minimum total area required. And it betrayed a faith in island biogeography theory that far exceeded what the evidence warranted. Together with the magnitude and pace of habitat destruction, the dubious status of island biogeography theory convinced many conservation scientists that the priority should be protecting species in the least area possible *now*.[5] Persistence, after all, *requires* representation: biodiversity must be adequately represented *before* its persistence can be ensured.

Despite these controversies, by the early 1980s one clear consensus had emerged: adequate reserve design required more data about species and ecosystems of conservation interest (see Gilpin and Diamond 1980; Higgs 1981; Simberloff and Abele 1982; Simberloff and Abele 1984; Willis 1984).

[5] MacArthur and Wilson's (1967, v) frank admission, "We do not seriously believe that the particular formulations advanced in the chapters to follow will fit for very long the exacting results of future empirical investigation," also supported this conviction.

Few doubted large areas were important for some species, such as large carnivores, and small areas adequately protected many species. The problem was that neither approach was *generally* defensible, and limited financial resources for conservation precluded following both. Before place-prioritization algorithms (PPAs), however, no methodology for reserve design existed that provided a defensible alternative to island biogeography theory.

But the theory, despite trenchant criticisms, significantly influenced the conservation community well into the 1980s (Kingsland 2002a, 2002b), especially in the United States. Although Diamond's (1975) principles were immediately criticized in high-profile journals like *Science*, the International Union for Conservation of Nature and Natural Resources (IUCN 1980), for instance, adopted them wholesale when devising a world conservation strategy. As presented there, the few qualifications of the principles Diamond (1975) originally made, and the criticisms of Simberloff and Abele (1976) and others were absent (Margules et al. 1982).

Given its influence and persistence despite incisive criticisms, many that were accepted by its original proponents (cf. Diamond 1976), the benefits the theory *would have* achieved should be appreciated. First, it would have obviated much of the need to acquire area-specific data. The urgency and limited funds for protecting threatened areas (Higgs and Usher 1980; Simberloff and Abele 1984), together with the high cost of data acquisition, made application of the theory to reserve design appear attractive and efficacious.

Second, as a widely publicized scientific theory, conservationists could invoke its authority to justify and convey an air of objectivity to their recommendations. This helped counter pro-development interests. Diamond (1976), for instance, emphasized the lack of a "firm basis" for predicting extinctions long before island biogeography theory. According to Diamond, this was a hindrance, "convincing government planners faced with conflicting land-use pressures of the need for large refuges" (1027). Since island biogeography theory might provide that basis, he argued, Simberloff and Abele's (1976) criticisms were troubling because "those indifferent to biological conservation may seize on Simberloff and Abele's report as scientific evidence that large refuges are not needed" (Diamond 1976, 1028). Terborgh (1976, 1029) seconded this worry: "Simberloff and Abele, if accepted uncritically, could be detrimental to efforts to protect endangered wildlife." Without the apparent scientific authority of island biogeography theory, the conservationist agenda seemed to lack justification: "pro-conservation individuals and groups, in and out of governments, hardly have a leg to stand on when competing for land and resources" (Soulé and Simberloff 1986, 44).

Controversy about the role of island biogeography theory in reserve design concerned more than its insufficient empirical support or inapplicability. Disagreement about the proper function of intuitions, presumably biologically informed in some way, also catalyzed controversy. In response to Simberloff and Abele (1976), for instance, Diamond (1976) emphasized that island biogeography theory may justify biologist's *intuitions* that many existing protected areas were too small to conserve their biota adequately. For Diamond, these intuitions likely track important truths and should thereby contribute to reserve design. Simberloff and Abele (1984) worried, however, that "theory is often seductive" (399) perhaps because of the often close connection between theory and intuition, and the contaminative effect the latter can sometimes have on the former. They instead held that inconclusive and speculative arguments, which are inadequate surrogates for field data, were nevertheless marshaling much of the perceived support for advocates of large conservation areas. As detailed below, PPAs and GIS provided a case-specific methodology that helped eliminate intuitive approaches to reserve design and transformed it into a characteristically data-driven science. Rather than trading in intuitions and theoretical implications that in fact are not, conservation biology came into its own as a science.

## 1.2 Politically Expedient "Worthless Land" Reserves

Science is a human activity, and largely a collective one. As such it is supported by and subject to the same cultural, economic, institutional, and political forces as any other human activity. Even researchers in so-called pure sciences must navigate the contours of these forces to secure grant funding and/or necessary permits, conduct human testing, get access to sensitive data, move lab spaces, and so on. But applied sciences in general, and certainly conservation biology in particular, are much more tightly tethered to those forces. This greater dependency has myriad consequences, obvious ones being more stringent scrutiny and regulation (e.g., strict safety standards for implantable medical devices, rigorous structural integrity testing for bridge designs). Another consequence affords an opportunity: the connection to broader societal concerns makes any advances in applied sciences all the more potent. Besides developing novel methods, theories, and discovering new truths, these achievements sometimes help solve the crucial societal problems applied sciences are designed to address. Or, as the case described below shows, progress can be made by demonstrating that a problem previously thought solved or goal previously achieved had in fact not been so.

The status of extant reserves was the issue. As an alternative to flawed theory-driven approaches to reserve design, Simberloff (1986) suggested returning to the intensive field research he thought characterized designation of national parks and reserves in the late nineteenth and early twentieth century. His belief, unfortunately, reflected the exception, not the norm. As debate about island biogeography theory and reserve design emerged in the mid-1970s, the ad hoc, unsystematic past bases for protecting particular areas, and the obstacles these posed for successful biodiversity conservation, became clear.

The historian Alfred Runte first made this criticism, about US parks. Borrowing California Senator John Conness's 1864 description of Yosemite Valley as "for all public purposes worthless" (quoted in Runte 1979, 48) in his argument it should have protected status, Runte (1972) proposed the "worthless lands" thesis. It claims that absence of certain kinds of economic value – primarily mineral and agricultural value – was the principal prerequisite for protecting areas. Scenic, recreational, or cultural values often constituted the publicized reason for protection, but the US government only seriously considered them when an area was clearly devoid of mineral or agricultural value. The thesis obviously bears on conservation. Since there is no reason to expect conservation value and mineral or agricultural value to be inversely correlated, or the former to be positively correlated with high scenic or recreational value, areas protected for these reasons probably do not achieve conservation objectives to any meaningful degree. If correct, the worthless lands thesis suggests these areas may be worthless to conservation.

Runte (1972, 1977, 1979) chronicled the ideological factors behind the creation of Yosemite (1864),[6] Yellowstone (1872), Mount Rainier (1899), Glacier (1910), Rocky Mountain (1910), Grand Canyon (1919), and other US national parks in support of the thesis. According to Runte, in the nineteenth century the United States lacked an internationally reputed literary or

[6] Legally, Yellowstone, not Yosemite, was the first national park. Originally Yosemite was transferred from federal jurisdiction to be managed as a state nature reserve by California.

artistic heritage, which prompted criticisms from European intellectuals.[7] One reason for creating national parks and celebrating their importance as part of US heritage, Runte argued, were feelings of cultural inferiority to Europe by US leaders. The natural wonders these parks protected were somehow taken to remedy this deficiency. For instance, when Clarence King, an early explorer and surveyor of the Sierra Nevada mountain range, considered the sequoia trees in 1864, he suggested that no "fragment of human work, broken pillar or sand-worn image half lifted over pathetic desert –none of these link the past and to-day with anything like the power of these monuments of living antiquity" (quoted in Runte 1977, 69).

Regardless of scenic or cultural value, however, the primary obstacle to protecting areas as national parks was farming, grazing, forestry, or mining potential. This was the US Congress's main contention in protecting Yellowstone, and it made very clear at the time that the Yellowstone Park Act (1872) would be repealed if commercial interests became apparent (Runte 1977). In Runte's estimation, the same noncommitment characterized much of the US government's conservation agenda in the 1960s and 1970s. In 1968, for example, Congress failed to protect a major watershed within Redwood National Park from deleterious erosion caused by adjacent logging. Similarly, its ostensible prohibition in 1976 of mining in Death Valley and other national parks actually sanctioned much of the strip mining that had motivated the public to demand prohibitory legislation. The historical precedent, Runte persuasively argued, was manifest: "ecological needs have come in poor second because the nation has been extremely reluctant to forego any reasonable opportunity, either present or future, to develop the national parks for their natural resources" (1983, 138).

The history of protected areas in Australia tells a similarly dispiriting story. Like the United States, Australian national parks and reserves were established where forestry, agriculture, mining, or commercial development were not viable (Recher 1976; Harris 1977; Hall 1988). For instance, the second legally designated national park after Yellowstone, Royal National Park (1879) outside Sydney, Australia, "was largely rugged, dissected sandstone plateau land, hopeless for agriculture and out of the way as far as Sydney Town of the

[7] For example, in 1820 the English clergyman Sydney Smith queried, "In the four quarters of the globe, who reads an American book? or goes to an American play? or looks at an American picture or statue?" (quoted in Runte 1977, 67).

1870s was concerned" (Strom 1979a, 46). Furthermore, its primary purpose was as a haven for recreation-seeking Sydney residents (Hall 1988). Other Australian protected areas, such as Flinders Chase National Park in western Kangaroo Island, were designated only after the government concluded they had little agricultural or mining potential (Harris 1977).

Another problem with the predominantly economic focus was that protected areas were susceptible to declassification if judged economically valuable later. A favorable turn in wheat markets and coincident increase in the value of wool due to US purchases for the Korean War motivated the Australian government to delist reserves for agricultural and pastoral development. From 1954 to 1962, 16,903 hectares were delisted in Australia (Harris 1977), a trend that continued well into the late 1980s (Hall 1988). Similarly, under the Reagan administration, in 1986 the Fish and Wildlife Service recommended large portions of Alaska's Arctic National Wildlife Refuge be explored for oil and natural gas, despite predicted adverse effects on wildlife (Tobin 1990).[8]

But blatant disregard for conservation priorities was only part of the problem. The lack of a method was another. Even after the Australian Fauna Protection Act (1949) was enacted, for example, absence of a systematic rationale for identifying areas to protect based on defensible conservation criteria prompted Strom (1979b, 68) to characterize the subsequent creation of parks and reserves as "a scramble for whatever was offering." Unsurprisingly, the areas that had been protected by the mid-1970s poorly represented major vegetation types and fauna inhabiting them (Recher 1976). In Tasmania, for instance, protected areas predominately sampled low economic value alpine areas and buttongrass plains ,while other important ecosystems went unrepresented (Hall 1988). The initially impressive increase in total protected area after the Fauna Protection Act – 53,947 hectares in 1937 to 249,260 hectares in 1954 – "must be tempered by the fact that in many cases the reasons for dedication had little or nothing to do with flora or fauna conservation" (Harris 1977, 63). Instead, areas were protected because, economically, it made little difference. Although politically expedient – politicians could vaingloriously publicize the truly substantive area they had helped set aside for conservation, all the while without any

[8] History repeats itself. The Trump administration pursued the same rapacious, transparently anticonservationist agenda for the Refuge.

real economic or political sacrifice – its cost was inadequate representation of Australia's species and ecosystems.

After the scope and severity of this problem was appreciated in the late 1970s and 1980s, conservationists still faced a serious challenge: demonstrating it beyond reasonable doubt. Doing so would preempt politicians from mollifying conservation demands by claiming that existing "worthless lands" protected areas achieved conservation goals. And, of course, conservationists also needed to identify, as precisely as possible, which areas should be protected for adequate representation of biodiversity. This would pressure politicians wishing to appear conservation-friendly to actually protect specific areas despite significant economic or political costs.

In combating powerful countervailing pressures, the lack of clear, quantitative assessments of the contribution (likely quite meager) that extant protected areas made to conservation goals, and what other areas could contribute, handicapped conservationists. Although such quantitative assessments do not guarantee governments would act accordingly, PPAs and GIS, which produced these assessments, provided clear information about the costs of forgoing conservation priorities. With these two technologies, which the increased speed and availability of microcomputers made much more readily accessible and usable, conservationists could employ this information to combat competing land-use interests more effectively.

## 2 An Algorithmic Turn in Applied Ecology: Place-Prioritization Algorithms

Section 3 of Chapter 5 surveyed the main features of PPAs in the course of critically analyzing the idea that they implicitly define biodiversity. Those details do not need rehearsing here. But understanding how PPAs helped overcome the obstacles described above, thereby supplying a compelling methodology for reserve design, requires recounting their emergence. The history is a case study in scientific progress.

Jamie Kirkpatrick, an Australian geographer with an interest in conservation, discovered the principle of complementarity at the core of PPAs in 1979 (Pressey 2002). While attempting to prioritize Crown-owned lands in Tasmania, Kirkpatrick first tried simple scoring methods. These methods involve attributing a quantitative score to each area in a region based on the number and kinds of species it contains. With this methodology, Kirkpatrick noticed that areas with several important species sometimes scored low, and several high-scoring areas sometimes shared almost all the same species. Prioritizing protecting areas based on score would potentially inefficiently over-protect several species while under-protecting others. To rectify this problem, Kirkpatrick formulated an iterative, heuristic scoring procedure. After selecting the highest scoring area, scores of unselected areas were recalculated on the assumption that the species in the highest scoring area were protected (i.e., complementarity). This ensured areas selected later complemented those selected earlier. Kirkpatrick's complementarity-based procedure was the first published PPA, appearing in a somewhat obscure report first (Kirkpatrick et al. 1980) and later in the only journal then devoted solely to conservation biology, *Biological Conservation* (Kirkpatrick 1983).

In the conservation context, the principle of complementarity was independently discovered three additional times: in the United Kingdom (Ackery and Vane-Wright 1984), Australia again (Margules et al. 1988), and South Africa (Rebelo and Siegfried 1990).[9] The first two complementarity-based PPAs were manually calculated, with pencil and paper, but by the late 1980s A. O. Nicholls utilized the increased availability and sophistication of microcomputers to program the first computerized PPA, presented in Margules et al. (1988). Besides heuristic PPAs, exact PPAs also originated in the 1980s in Australia. Cocks and Baird (1989) first utilized a commercial integer programming software package to identify optimally efficient reserves. They prioritized areas within the Australian Eyre Peninsula according to several representation targets. Advances in microcomputing throughout the 1980s made this computationally intensive analysis, virtually impossible a decade earlier, feasible.

From the outset, the importance of PPAs to reserve design was abundantly clear to its developers and most of the community of applied ecologists working on conservation issues.[10] This helped shift the goal from preserving maximum species richness at some hypothesized future equilibrium level as putatively counseled by island biogeography theory (see Section 1 above) to

[9] See Justus and Sarkar (2002) for a detailed history.

[10] This was not the case in the United States. For example, the first conservation biology textbooks, written by US conservation biologists, did not discuss PPAs at all (Primack 1993; Meffe and Carroll 1994). Much of their discussion of reserve design focused on Diamond's (1975) principles, whose epistemic credentials had aged quite poorly over almost two decades. Meffe and Carroll (1994), for instance, incredibly claimed the species–area relationship justified large conservation areas over small ones.

representing biodiversity surrogates now. In terms of methodology, it refocused attention from general theories to algorithmic procedures that require geographically explicit data. By producing complementary sets of areas achieving representation targets, PPAs ensure biodiversity is represented efficiently in the smallest total area possible, which is imperative given the limited monetary resources available for conservation. As Margules et al. (1988) emphasized about the SLOSS debate, conservation biologists had excessively focused on ecological processes within potential reserves without first adequately understanding how to represent biodiversity in protected areas efficiently. PPAs rectified that deficiency.

In so doing, PPAs filled the methodological hole left by the failings of reserve design principles supposedly based on island biogeography theory. This transition was on clear display in a special issue of *Biological Conservation* a year after Margules et al. (1988) proposed the first computer-based PPA. The issue's overarching topic was the growing trend toward computer-based methods, led by Australian conservation scientists at the New South Wales National Parks and Wildlife Service (NSW-NPWS) and Australia's Commonwealth Scientific and Industrial Research Organisation (CSIRO). In its introduction, Margules (1989a) indicated that none of the papers addressed the long-term adequacy of reserves, for two reasons. First, he emphasized, processes leading to extinction were poorly understood. Despite claims made on behalf of island biogeography theory, claims contrary to its originally exploratory nature, long-term field studies required for a better understanding of extinction had yielded little helpful insight. The second reason constituted the conceptual basis for PPAs:

> knowledge of patterns of species distributions has priority over a knowledge of ecological processes in our efforts to maintain biological diversity. Techniques for managing reserve systems to prevent extinctions will not maintain diversity if the reserve systems being managed do not contain the full range of species in the first place. (8)

As a data-driven methodology, PPAs supplied the scientific basis for reserve network design that island biogeography theory could not. Its focus on geographically referenced data, not hypothesized equilibria derived from controversial theory, and its mechanistic application of explicit and relatively uncontroversial conservation criteria made results of place prioritization compelling.

Based on his analysis of Crown lands in Tasmania, for example, policy makers acted on each of Kirkpatrick's (1983) seven recommendations for new conservation areas (Pressey 2002; see Figure 8), and achieved unprecedented success.

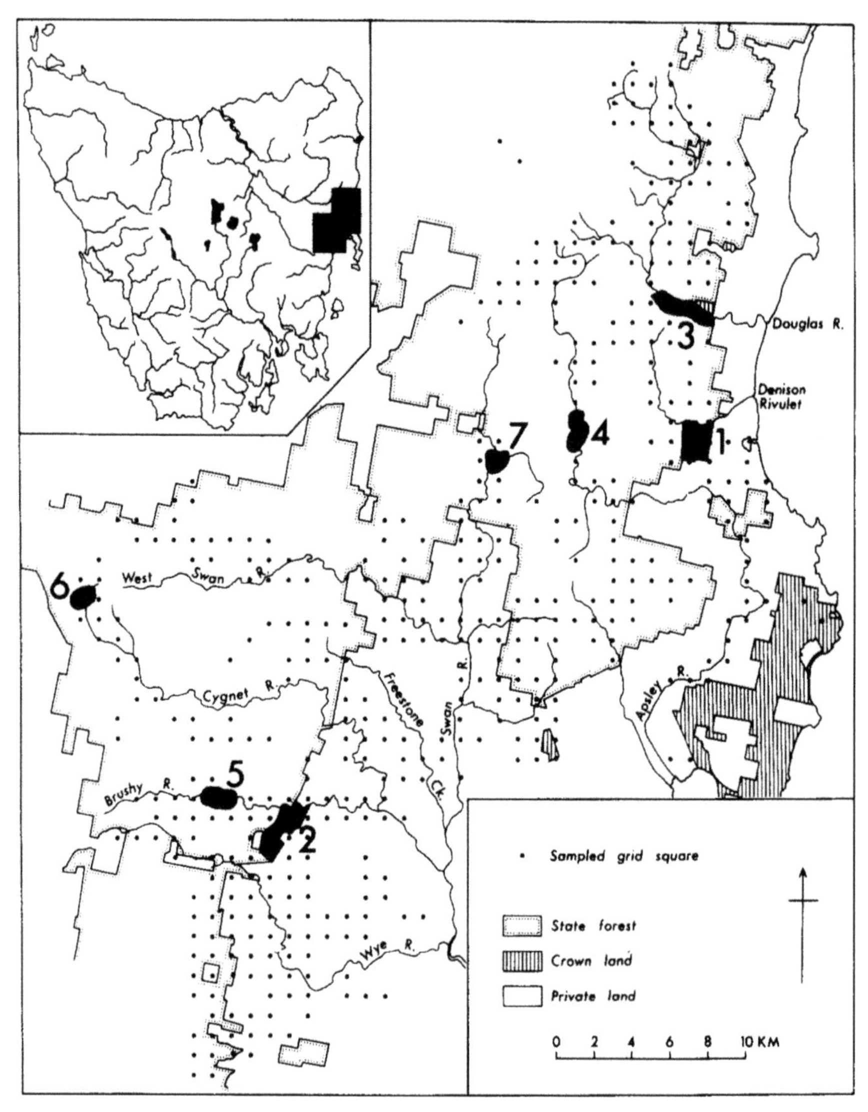


Figure 8 Kirkpatrick's (1983) seven recommendations for new conservation areas in Tasmania based on the first place-prioritization algorithm. The number labels indicate the ordinal priority of each shaded area. From Kirkpatrick 1983, 133

This stood in stark contrast to Australian politicians' legacy of ignoring the counsel of informed conservation interests inside and outside government (Harris 1977). The explicitness of his algorithm helps explain its power. In an interview, Kirkpatrick added that its effectiveness was due to "the desire of the forestry people to appear scientific in their conservation efforts ... the logic of the process, and its minimalism, also appealed" (quoted in Pressey 2002, 436). Correct perception of the algorithm as a repeatable, objective process based on sound data made policy makers more receptive to the priorities it yielded, especially as a scientifically grounded alternative to more economically threatening agendas of other conservation groups in Tasmania (J. Kirkpatrick, personal communication).

With such a different data-driven approach to reserve design, it is unsurprising that results of prioritization often conflicted with the intuitions motivating earlier, theoretically oriented strategies. In general, complementarity conflicted with Ehrenfeld's (1976, 653) assertion that "The need to conserve a particular community or species must be judged independently of the need to conserve anything else," and the actual prioritizations PPAs produced revealed similar conflicts. Margules et al. (1988), for instance, prioritized areas within the Macleay River floodplain of northern New South Wales with respect to two targets: (1) representing ninety-eight native plant species and (2) representing these species and nine wetland habitats. The two PPAs they implemented for each target selected 44.9% and 75.3% of the region analyzed, respectively. Most importantly, similar to Kirkpatrick's (1983) analysis, the pattern of selected areas achieving the targets did not fit any of Diamond's (1975) reserve design principles (Robert Pressey, personal communication). Unlike island biogeography theory, reserve design based on algorithmic prioritization depends essentially on the specific locations of biodiversity features, in this case species and habitats. Their analysis therefore confirmed with hard data the general suspicion that reserve design principles that fail to consider this kind of geographically explicit information would usually yield poorly representative reserves (see Section 1). Protecting a few large areas in the Macleay River floodplain might protect many species and habitats, but not all ninety-eight native species and nine wetland types, nor would those protected be protected in the smallest area possible.

Besides freeing reserve design from the grip of island biogeography theory, one essay in the special issue also demonstrated how PPAs could verify the "worthless lands" thesis, by revealing with quantitative precision how poorly existing reserves represented biodiversity.[11] With a modified version of an algorithm of Margules et al. (1988), they prioritized areas in western New South Wales for targets of one and five representations of 128 "land systems." Somewhat similar to habitat types, land systems were classified according to topography, soil, and vegetation type. Their analysis quantitatively confirmed what history suggested (see Section 1): the thirteen existing conservation areas in the study region poorly represented the 128 land systems. Pressey and Nicholls (1989) conducted four prioritizations: (1) two with the two targets in which the thirteen existing conservation areas were assumed protected and the land systems they contained already protected and (2) two with the same targets in which the thirteen areas were not assumed protected. To achieve the first target, 11,503 square kilometers were required if the thirteen areas were assumed protected, compared with 7,980 square kilometers if they were not. The five-representation target required 30,065 square kilometers if the thirteen areas were assumed protected, 28,726 square kilometers if they were not. The additional area required for the one-representation target was 44% of the total area required when the thirteen conservation areas were not assumed protected, a dramatic indication of substantial inefficiency. This was especially problematic since the total area required for the one representation target, let alone the five, when the thirteen areas were excluded exceeded all reasonable estimates of the financial resources available to protect areas. Every conservation dollar counts, and extant protected areas wasted them.

Pressey and Nicholls (1989) also noticed a second negative aspect of the thirteen already protected areas. Besides poorly representing land systems, they also hindered efficient construction of a fully representative reserve. Even if areas complementary to the thirteen conservation areas were protected, the resulting reserve would still achieve representation targets much less efficiently than a reserve constructed from scratch. Delisting some or all of these thirteen areas might therefore improve efficiency of future reserve design in New South Wales.[12]

[11] Margules (1989b) and Pressey and Nicholls (1989) share the distinction of being the first to do this.

[12] Pressey and Nicholls (1989) did not recommendation this, probably for fear of how it could be used in political contexts.

Margules (1989b) reached similar conclusions about conservation areas in the Mallee area of South Australia. Using a slightly modified version of an algorithm from Margules et al. (1988), Margules (1989b) found that only six of the eighteen areas required to achieve one representation of forty-five vegetation "alliances" were included in existing reserves. Hence, fourteen of the twenty-one conservation areas in the region were superfluous for this representation target. Furthermore, if these twenty-one areas were assumed protected, 82% of the region was required to represent each alliance, compared with 69% if the reserve system was constructed from scratch.

In their prioritization of the Cape Floristic Region of South Africa to protect 326 taxa (species and subspecies) of fynbos vascular plants, Rebelo and Siegfried (1990) showed that less than 27% of existing conservation areas and 50% of areas proposed for protection included areas of high fynbos endemism. Furthermore, they suggested that De Hoop Nature Reserve, which composed 55% of the Cape's total protected area, was in its least diverse part and should be delisted in exchange for new areas containing more fynbos plants. Although conservation biologists had considered delisting reserves before (Siegfried 1978, 1989), this was one of the first quantitative justifications for delistment, based on a PPA. The "worthless lands" status of reserves could no longer be simply ignored by the powers that be.

Throughout the 1990s an Australian conservation biologist, Robert Pressey, publicized the capability of PPAs to quantify deficiencies of existing reserves objectively (Pressey 1990a, 1990b, 1992, 1993, 1994; Pressey et al. 1990; Pressey and Tulley 1994). He also recognized that the speed and flexibility of PPAs, especially heuristic PPAs at the time, would make them powerful tools in policy-making contexts. First, heuristic PPAs could rapidly and quantitatively demonstrate the requirements of various representation targets. This helped expose politicians' tepid commitment to conservation, and thwart the pattern of "politically driven reservations [that] are often no more than environmental gestures of real concern" (Pressey 1992, 20). Place prioritizations quantified the often severe conservation costs of land-use policies, which could then be conveyed to the public.

Second, PPAs were flexible enough to implement several different kinds of constraints on place prioritization. For example, PPAs could explicitly exclude inappropriate areas or mandate inclusion of others, as well as incorporate other criteria, such as adjacency or compactness, into prioritization (see Margules and Sarkar 2007). The adjacency criterion, first integrated into a PPA by Nicholls and Margules (1993), prioritizes adjacent areas over nonadjacent ones and tends to generate reserve systems with areas clustered around a few locations. The criticism, then, that "complementarity-based algorithms are biased in favor of selecting small, potentially isolated reserves containing relatively low species abundances" (Linquist 2008, 541) is misinformed. PPAs are neutral on reserve size in two ways. First, the degree of "clumping" of selected cells can be intentionally varied with adjacency, compactness, and other selection criteria in PPAs. It is entirely up to the algorithm user to specify the appropriate degree, which certainly can be quite high. Second, PPAs are neutral on cell size. The algorithm user determines their size when gridding the region of interest into cells. The size can be very large, and often is to accommodate the habitat requirements of some species. And if abundance data are available, prioritization can utilize that information in a variety of ways.

The criticism that "complementarity-based algorithms overlook the habitat requirements of particular species" (Linquist 2008, 542) similarly shortchanges PPAs. The nature of the claimed problem is clarified: "A complementarity approach might prioritize areas containing high levels of frog diversity, but inadvertently exclude suitable breeding habitats where these animals are rarely found. Although island biogeography theory does not take particular habitat requirements into account, its preference for large continuous land reserves errs on the side of caution" (542). With respect to the first sentence, it "might" but it certainly need not. If knowledge of that habitat exists, the breeding sites can easily be included in any prioritization. In fact, their selection can be *mandated*. PPAs are very flexible tools that can integrate many different kinds of information. They are regularly integrated with species distribution models that would typically capture these habitat relations (see the next section). If there is any potential problem here, it is a dearth of information. With respect to the second sentence, besides the blindness of island biogeography theory to habitat requirements, it is quite uncertain what caution counsels given a finite budget for land acquisition. Large area *here* means smaller and/or fewer areas *elsewhere*. Absent distributional data, which scenario is superior is utterly unclear. In general, PPAs are tools that complement and can integrate, rather than compete with, sound input from ecological science.[13]

Given their speed and flexibility, conservation biologists could use PPAs to revise conservation planning strategies rapidly when new information about the analyzed region becomes available or when politicians want to consider different representation targets (Pressey and Nicholls 1989). Although heuristic PPAs did not guarantee optimally efficient reserves as exact PPAs did, a weakness first pointed out by Cocks and Baird (1989), in the 1990s the computation time the latter required slowed or, worse, prevented effective conservation planning altogether, especially for analyses of large regions (Pressey and Tulley 1994). This explains the early predominance of heuristic PPAs over exact PPAs in real-world conservation planning.

In policy-making contexts, these features of PPAs made negotiations between conservation and competing interests more explicit and quantitatively rigorous, especially when combined with the analytic and computational capabilities of GIS. This combination, in turn, resulted in some unprecedented conservation successes. Before describing them, the emergence of GIS as a powerful tool within applied (and "pure") ecology needs to be told.

## 3 The Algorithmic Turn II: Geographic Information Systems

Geographical information systems (GIS) are computerized systems that facilitate integration, analysis, and visualization of different kinds of geographically referenced data. In GIS, for example, precipitation, temperature, vegetation cover, soil type, species distributions, land ownership, and human population data for an area can be visualized and analyzed at different spatial scales.[14] These data can be derived from numerous sources, including remote sensing by satellites.

[13] A final supposed problem is worth briefly addressing. Linquist (2008, 542) says, "A final shortcoming of the complementarity approach is that it is sensitive to 'apparent novelties' or species that appear rare in a region but occur in large numbers outside the area being investigated." But as common sense dictates, such species should, and almost surely would, be excluded from the place-prioritization analysis, absent some overriding reason, such as a desire to protect bald eagles in the continental United States as a national symbol even given abundant populations in Canada and Alaska. Nothing about PPAs forces the hand of their users. It is up to them to make the obvious exclusion.

[14] A complete description of GIS's analytic capabilities, even in the late 1980s, is impossible here. For such a description in 1987, see Smith et al. (1987).

Before GIS had any impact on conservation biology, it revolutionized geography and cartography. GIS helped overcome two limitations of physical maps: (1) their limited information storage capabilities and (2) the extremely complicated and time-intensive calculations they require to retrieve and manipulate this information. Roger Tomlinson, director of the Canada Geographic Information System from 1960 to 1969, estimated that most measurement operations on 10-square-inch physical maps require 80–650 minutes to complete (Tomlinson 1988). Within GIS, these operations required a matter of minutes; today they usually require seconds or fractions thereof. The widespread dissemination of personal computers in the late 1980s also made GIS much more readily available. Many sophisticated statistical and mathematical operations that were effectively impossible to perform on physical maps could be performed easily by GIS on personal computers (Abler 1987).

In his defense of the discipline of geography, Ronald Abler, the president of the American Association of Geographers, remarked that "GIS technology is to geographic description and analysis what the microscope, the telescope, and computers have been to other sciences" (1987, 514).[15] Concurrent to its revitalization of geography, in the mid-1980s Australian and US applied ecologists recognized the role GIS could play in the analysis and storage of geographically referenced data for reserve design.

In Australia, research groups at CSIRO and NSW-NPWS began developing simple GIS for general land-use planning. One of the first was the LUPLAN computer package (Ive and Cocks 1983), development of which began in 1978 at CSIRO to automate a land-use planning method, SIRO-PLAN (Cocks et al. 1983; Ive et al. 1985). For the region studied, LUPLAN could digitally store attribute data, such as natural resource distributions, soil and vegetation types, present and past land use, and geological properties. LUPLAN then generated ratings of specific areas in the region based on their contribution to different land-use options such as conservation, recreation, grazing, or extractive industry. With these ratings, different land-use policies could be evaluated quantitatively, and LUPLAN produced visual representations of the evaluations.

[15] Abler's presidential address followed the dissolution or severe reduction of geography departments at the University of Chicago and Columbia University in 1986, and Northwestern in 1987.

According to Cocks and Ive (1988), two virtues of LUPLAN were its ability to assess land-use policies quantitatively and rapidly, and to accommodate changes in land-use options and attribute data in its assessments. The latter provided the flexibility required in policy-making contexts, and the former helped "provide information which allows [land-use policies] to be defended and debated in substantive rather than emotional terms" (Cocks and Ive 1988, 258). Another virtue was that attribute data acquired in the past could be stored by LUPLAN, and LUPLAN could integrate new data into the database easily. Although LUPLAN was originally programmed for 48-kilobyte microcomputers (Cocks and Ive 1988), increases in computing speed and data storage capacity throughout the 1980s ensured it and other GIS could store and analyze an effectively limitless amount of data. In 1985, Simon Ferrier began developing a more sophisticated GIS, the Environmental Resource Mapping System (E-RMS), at NSW-NPWS. From its completion in 1988 (Ferrier 1988) to its replacement at NSW-NPWS by ArcView in 1997, E-RMS was used extensively by NSW-NPWS and more than a hundred organizations throughout Australia.

E-RMS and LUPLAN helped resolve a formidable obstacle to successful reserve design: absence of adequate data on the distributions of biodiversity surrogates. The problem had two aspects: (1) for several regions such distributional data did not exist, and (2) existing data "have often been acquired without the guidance of a co-ordinated information management goal, are scattered among different institutions in incompatible formats, and, as such, are often difficult to locate" (Davis et al. 1990, 55).

Davis et al. (1990) first described the capability of GIS to resolve problem (2). They outlined development of a comprehensive GIS-based biodiversity information system that would synthesize environmental data already collected for limited purposes. Specifically, different kinds of data sets developed with different data collection methodologies and for different purposes could be integrated into a single database. Applied ecologists could then utilize a single GIS software package to analyze relationships between a wide variety of different kinds of attributes relevant to conservation planning: species, vegetation and soil types, elevation, precipitation, land use, resource composition, human density, predicted human population trends, and so on. Data collected at different spatial scales could also be easily integrated.

Ferrier realized GIS could help resolve problem (1) by assisting in field survey design and survey data analysis (Ferrier 1990; Ferrier and Smith

1990). Part of the problem was that distributions of many biodiversity surrogates, for instance, animal species, could not be predicted reliably from remotely sensed data and were reliably obtainable only from exhaustive, time-consuming, and expensive field surveys. Since such surveys usually exceed funds available for data acquisition, conservation biologists focused on maximizing the information obtained from limited surveys. One way to do this was to stratify survey sites with respect to environmental gradients. This ensures extrapolation to unsurveyed sites is as justified, and informative, as possible.

Before GIS, surveys were usually stratified along a single environmental gradient, such as vegetation type (Ferrier and Smith 1990). This methodology unrealistically assumed a single environmental parameter predominantly influences patterns of species distributions and other components of biodiversity. GIS's ability to analyze relationships between different environmental parameters rapidly helped design surveys stratified with respect to numerous gradients. This kind of analysis was based on two types of database queries: (1) What are the attributes of a set of locations? and (2) What locations possess a set of attributes (Davis et al. 1990)? With query (1), for instance, different kinds of data could be "overlaid" to determine an area's species composition, correlations between species distributions and environmental parameters, threats to species from human population growth trends, and so on.[16] This helped identify survey sites simultaneously stratified to several environmental gradients (Ferrier and Smith 1990). The result was more representative and, therefore, more informative surveys. Researchers in the Department of Ecosystem Management at the University of New England Armidale and NSW-NPWS implemented this multivariate survey methodology with E-RMS in faunal surveys in northeastern New South Wales in the late 1980s (Smith et al. 1989).

GIS also facilitated multivariate analysis of correlations between survey data and remotely sensed data, which was the basis for developing predictive models of species distributions to extrapolate across unsurveyed areas (Ferrier and Smith 1990, figure 5). Such models existed before GIS use became widespread (e.g., BIOCLIM; see Busby 1991), but without integration into GIS, spatial analysis of relationships between biological and

[16] McHarg (1969) originally pioneered this overlay method without the assistance of a computer in the context of ecologically sensitive urban planning.

environmental variables was difficult. To surmount this difficulty, from 1988 to 1991 Ferrier's research group at NSW-NPWS added a module to E-RMS that predicted species distributions from environmental data (Ferrier 1992). In this way, GIS facilitated the marriage of species distribution models developed by ecologists with the data those models required to make confirmable predictions.

Admittedly, predicted species distributions are not substitutes for ground-truthed survey data. GIS-based predictive models, however, also helped identify which areas should be more systematically surveyed and/or take initial priority in reserve design. If conservation action was urgent, as it often was, GIS-based models could rapidly predict species distributions in a region. Conservation biologists could then identify potential biodiversity hotspots and rapidly ground-truth these predictions or, if immediate action was necessary, recommend they be protected. By generating these predictions, GIS-based predictive models partially circumvented the perceived need for theoretical approaches to reserve design motivated by a recognized absence of adequate data. GIS (and PPAs) are "data-hungry," but in reserve design there is no methodological alternative to data but blindness, and the technologies themselves furnished the means for acquiring new data, enhancing the inferential power of extant data, and mitigating the effects of paucity of data.

Similar to PPAs, GIS-based models predicted species distributions that also showed that Diamond's (1975) principles would produce poorly representative reserves. In an analysis of these principles, Margules and Stein (1989) used a rudimentary GIS and model of thirty-two *Eucalyptus* tree species to evaluate seven reserves in coastal and sub-montane areas of southeastern New South Wales. Predicted distributions of the thirty-two species demonstrated only one of forty-three simulated large reserves with the same total area as the seven existing reserves would contain as many *Eucalyptus* species; that is, designing reserves after principle B from Diamond (1975) would poorly protect them. Furthermore, the only exception was an elongated rectangular area spanning gradients in rainfall and temperature. Thus, any reserve conforming to Diamond's (1975) principle F would contain fewer *Eucalyptus* species. These results reinforced Simberloff and Abele's (1984, 400) criticism that, "For the minority of systems where shape is significant, it is just as likely that long thin islands have more species than do round ones as vice-versa." More importantly,

Margules and Stein's (1989) analysis confirmed the need for reserve design based on data about the specific characteristics of regions: "The current distributions of species form the most rational basis for reserve selection, not theoretical ideas on reserve size and shape" (221).

In the first publication conveying the importance of GIS to conservation biology, Scott et al. (1987) showed that its ability to analyze different kinds of data alone provided a simple method for evaluating existing reserves. They recounted one of the earliest GIS-based analyses of a reserve system (Kepler and Scott 1985) to illustrate the methodology. With a GIS, Kepler and Scott (1985) had analyzed reserves in Hawaii with distribution and population density data for endangered bird species, and vegetation distribution data. The analysis consisted of overlaying these kinds of attribute data with the geographical pattern of the existing Hawaiian reserve system (Figure 9).

High degrees of overlap between existing reserves, species distributions, and the vegetation types known to be important parts of their habitats would show the reserves were representative. Similar to PPAs, however, Kepler and Scott (1985) found few high-richness areas and fewer areas of high population density of endangered bird species in the Hawaiian reserves.[17]

The disparity revealed by GIS between the extant reserves and endangered bird distributions confirmed the drawbacks of the non-conservation rationale underlying the former. These specific locations were originally protected as part of anthropogenically important watersheds, or for having unique geological features, such as volcanic activity (J. Michael Scott, personal communication). The GIS analysis did, however, have a positive impact. It identified new conservation sites that more adequately represented these species, which prompted creation of the 6,693-hectare Hakalau Forest National Wildlife Refuge (Scott et al. 1993).

More generally and importantly, as a reserve design methodology, GIS analyses "led to results different from the conventional wisdom" (Scott et al. 1987, 785) and, presumably, the intuitions underlying that "wisdom." Contrary to assurances from politicians, the GIS analysis showed that the

[17] In fact, J. Michael Scott estimated that the overlap between the areal extent of the reserves and the distributions of endangered bird species was only 10% (J. Michael Scott, personal communication).

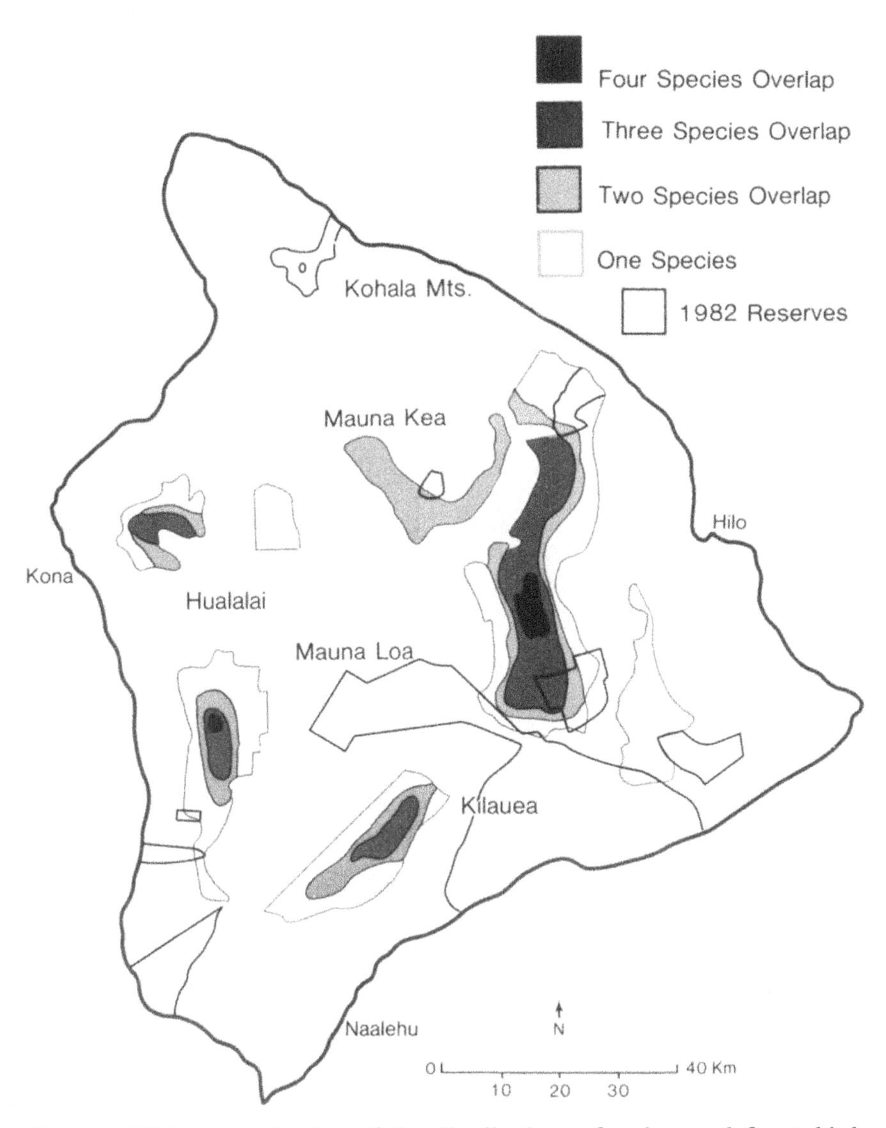


Figure 9 GIS-based evaluation of the distributions of endangered forest bird species and existing reserve system on the Island of Hawaii in 1982. From Scott et al. 1987, 785

extent of area protected in Hawaii was woefully short of that necessary for at least avian conservation. In addition, there was not, like the results of the analysis of Margules et al. (1988), any similarity between the candidate reserves suggested by the GIS analysis for the Hawaiian bird species and Diamond's (1975) principles. This methodology for identifying deficiencies in existing reserves and priorities for protecting new areas was labeled "gap analysis" by Burley (1988) and later elaborated in detail by J. Michael Scott and others (Scott et al. 1991a, 1991b, 1993).

By the early 1990s, it was clear that narrowly focusing conservation efforts on endangered species and small populations was problematic and that broader approaches to biodiversity conservation were needed (see Chapter 5). As Scott et al. (1987, 783) put it, "it is a sad commentary that the current widespread practice of 'Emergency Room Conservation' channels most of the economic and emotional support for the protection of biological diversity into those few species least likely to benefit from it." The problem was that, in the United States, the Endangered Species Act (1973), National Forest Management Act (1976), and other legislation, though legally powerful, focused conservation strategies and resources on small populations; minimum viable population analysis and population viability analysis are prominent examples. This significantly diminished the limited financial resources available for more general biodiversity conservation. In 1985, for example, $1.25 million was spent preserving nine California condors (McKendry and Machlis 1991), a sum exceeding the annual expenses of an average several-thousand-hectare conservation area protecting hundreds or thousands of species (Scott et al. 1987). GIS-based gap analysis not only provided a practical methodology for addressing general biodiversity conservation priorities, but it did so cheaply and generated results rapidly. Subsequently, many conservation agencies quickly adopted the methodology. Following a pilot study started in Idaho in 1987 (Scott et al. 1993), by 1991 at least thirteen US states were conducting GIS-based gap analyses (McKendry and Machlis 1991), and by 1994 thirty-two states were doing so (Csuti 1994).

For all its merits, GIS-based gap analysis had one important disadvantage: by identifying areas of high species richness or predicted high richness as potential conservation areas, it failed to ensure complementarity of recommended reserves. The resulting reserve system, therefore, would usually not protect the most species in the smallest total area. This was rectified by integrating PPAs and GIS into a single computerized system. In collaboration with conservation biologists at the University of Idaho, geographers at Clark University were some of the first to develop such a system (McKendry and Machlis 1991). They utilized a modified version of an algorithm of Margules et al. (1988) to prioritize 635-square-kilometer hexagonal areas in Idaho to ensure several native vertebrate species were represented at least once.

Combining PPAs and GIS proved to be a powerful tool. On April 23, 1996, negotiations began between government, industry, and conservation agencies regarding 2.4 million hectares of forest in eastern New South Wales (Pressey 1998). Negotiations were facilitated by a decision support system that integrated C-Plan ("C" for "conservation"), a heuristic PPA developed by Robert Pressey and Mathew Watts (Pressey et al. 1994), and E-RMS (RACAC 1996). After four weeks, representatives reached an unprecedented agreement, designating nine new reserves and national parks (250,000 total hectares) and deferring logging of 816,000 hectares of forest.[18]

This astounding success depended on the transparency and flexibility of the C-Plan/E-RMS decision support system. As Tom Barrett of NSW-NPWS, who managed the system during negotiations, put it, "The idea is to make the whole process as scientific and transparent as possible" (quoted in Finkel 1998a, 1789). The explicitness of C-Plan allowed representatives of all interests to see why particular areas were selected to achieve specified representation targets. E-RMS facilitated visualization of different conservation scenarios generated for different representation targets and rapid query of the data generating them. The C-Plan/E-RMS system was also flexible enough to analyze several variations in representation targets and prioritization constraints imposed by negotiators. These virtues, of course, were effective only because analyses were executable in seconds or at most minutes, and their results could then serve as bases for compromise and reassessment (Pressey 1998). Without the ability of E-RMS to integrate and rapidly analyze large amounts of different kinds of data, and without the ability of C-Plan to prioritize areas for different representation targets rapidly, these negotiations and their impressive results would have been impossible.

The power of combining PPAs and GIS was recognized not only by Australian conservation biologists. By 2000, heuristic PPAs alone had been used in academic studies in at least ten countries and in policy-making contexts in six countries (Justus and Sarkar 2002). Today there is consensus: PPA and GIS are indispensable tools of reserve design.

[18] Unfortunately, similar negotiations in 1997 and 1998 were not as successful for political reasons (see Finkel 1998b; Justus and Sarkar 2002). Even so, C-plan and E-RMS supplied assessments by which conservation biologists could quantify their failure (and success) and communicate it as such to the scientific community and broader public.

## 4 "Progress" by Any Other Name

In one of the new discipline's first manifestos, Soulé (1985) deemed conservation biology a "crisis" discipline with an explicit normative agenda: preserving biodiversity. Providing a scientific basis for efforts to do so was (and is) the overarching goal, and this substantially shaped its early aspirations, priorities, and real-world conservation strategies. But as the preceding discussion makes clear, without two technological innovations – place-prioritization algorithms and geographical information systems – supplying that basis, and its emergence as a *rigorous science* in general, would have been impossible.

Despite the impressive credentials of the advances recounted above, that they constitute real "progress" has been called into question. Linquist (2008, 531) is skeptical: "[T]he adoption of complementarity-based algorithms in place of theoretically motivated conservation guidelines has arguably not advanced the field of conservation biology." Some misconceptions about PPAs motivate this judgment (see Section 2 above), as well as an apparent underappreciation of how badly Diamond's design principles fare empirically (see Sections 2 and 3 above). But the core of the criticism seems to be something else, a view of what scientific progress should look like. After noting how conservation biologists have largely abandoned Diamond's principles, rightly given the failings discussed above, Linguist (2008, 530) admonishes, "But instead of rigorous autoecological studies, *something much less scientifically grounded* emerged in their place. Conservation biology has become dominated by various 'fast and frugal' place prioritization algorithms for designing conservation reserves" (emphasis added). As the juxtaposition insinuates, PPAs comparatively lack a scientific grounding that island biogeography at least could have, even if numerous empirical studies suggest it does not.

There are two interrelated aspects of this contention. The first is the idea that "the core principle on which most place prioritization algorithms are based – the principle of complementarity – is not ecologically sound" (Linquist 2008, 531). It is hard to understand the motivation behind this claim. "Complement" in "complementarity" is set-theoretic. In the course of a prioritization, the biodiversity surrogates in already selected cells are assumed to be protected. Selecting by complementarity requires selecting the cell that maximizes surrogate representation in the complement of biodiversity surrogates not yet assumed protected. Since the goal is protecting as much as possible with the usually severely limited resources available to do so, the principle of complementarity simply expresses how set theoretic facts dictate what is instrumentally rational given this goal. The principle is not an ecological one whose soundness depends on findings in ecological science any more than the formula for a mean is biological and subject to evolutionary analysis when employed in studies of population genetics.

A second aspect concerns a more widespread view of how scientific progress should be conceptualized. Bird (2007), for example, countenances three approaches to characterizing scientific progress: an epistemic approach that gauges progress in terms of acquiring knowledge, a semantic approach that gauges progress in terms of scientific theories being nearer the truth, and a functional-internalist approach that gauges progress in terms of "problem-solving" à la Kuhn. Bird comes down decisively in favor of the epistemic approach, but it (and the semantic approach) seems utterly ill-equipped to account for progress in applied, ethically driven sciences. These sciences do not deliver anything resembling justified truth belief about the cosmos. Instead, they provide a scientific basis and scientifically derived means for achieving ethical goals. It seems that only a functional account, appropriately stripped of unnecessary Kuhnian trappings, can account for these achievements being genuine progress.

Bird's main charge against the functionalist approach is that it is unintuitive – "puzzle solving" by false theories does not intuitively constitute scientific progress (Bird 2007, 83) – and that Kuhn's view of the nature of science is flawed. The latter can be accepted; the former should be rejected. Trading in intuitions is, of course, a game for those who believe they have significant epistemic weight. But it is worth emphasizing that the relevant intuition is indefensible, especially for ethically driven sciences. First, note that what generates the unintuitiveness is the theory's falsity. Anti-realists unencumbered by such semantic preoccupations are accordingly free to evaluate whether the work being done by theories constitutes progress on other, less obscure grounds. Second, truth values and theory status are ancillary to judgments of progress for applied, ethically driven sciences. What really matters is whether the insights and tools the science delivers are promoting the goal's achievement. It is conceivable, I speculate, that a grand, unified, true theory of biodiversity could emerge in the distant future. That theory would be an amazing achievement in biological science, rivaled perhaps only by Darwin. But its progressive import *in conservation biology* would be measured by how it helped move forward the agenda of conserving biodiversity.[19]

In fact, Bird endorses this broader understanding of progress elsewhere in his analysis. He says (2007, 83):

> Our conception of scientific progress is linked to what we take the aim of science to be. In general, something like the following principle holds:
>
> (A) if the aim of X is Y, then X makes progress when X achieves Y or promotes the achievement of Y.

Precisely. With respect to whether (A) holds when Y = biodiversity conservation and X = place-prioritization algorithms and geographical information systems, let's review. Within roughly two decades, PPAs and GIS transformed reserve design from a problematically intuition- and theory-driven affair into a data-driven, quantitative science. Specifically, these technologies demonstrated why conservation biologists should abandon principles "inspired" by island biogeography theory, and they provided a defensible alternative, a methodology, moreover, that maximizes representational efficiency given limited budgets for designating reserves. This methodology also produced quantitative (usually critical) assessments of existing "reserves" that helped expose Machiavellian subterfuge about being conservation-friendly by politicians and other power brokers. Finally, these technologies provided area-specific reserve design recommendations and rapid assessments of proposals in policy-making contexts that led to unmitigated and unprecedented conservation successes. If this does not constitute progress in an applied science, nothing does.

[19] It goes without saying that achieving the goals of ethically driven sciences – improving human health, increasing economic well-being, fostering cooperation and sociality, conserving biodiversity – are hardly akin to "puzzle-solving."

# 7 Fact and Value in Applied Ecology

This last chapter recounts one of the crowning success stories of applied ecology, the development of place-prioritization algorithms and geographical information systems. The advances these algorithmic technologies afford did not emerge from penetrating theoretical insights that catapulted theory construction forward or from crucial experiments with similar epistemic significance, as is so often heralded in histories of other sciences.[1] Rather, progress was made via the more effective marshaling of data: integrating different kinds of georeferenced data, better extracting salient information from that data, pinpointing what new data are most needed, and so on. This progress also was not measured in typical metrics of scientific success, such as proximity to truth, explanatory power, or theoretical unification (although these can obviously be relevant). Instead, the appropriate metric is specific, pragmatic, and ethical: providing a scientific basis for biodiversity conservation. For applied sciences with ethical objectives like conservation biology, a defensible conception of progress must be a broad, ethically value-laden one, and it certainly need not be theory-based.

But this ethically laden character broaches an important issue. What is the status and *proper* role of nonepistemic values in such goal-oriented sciences, which besides conservation biology include restoration ecology, invasion biology, agricultural science, climate science, medical science, clinical psychology, welfare economics, and many more? Whereas most so-called basic sciences are principally concerned with discovering and explaining phenomena, applied sciences often have a different, explicitly teleological agenda to pursue more immediately pressing goals. Applied ecology devoted

[1] See Kuhn (1962, 1977) for a penetrating critique of the veracity of many such histories.

to conserving biodiversity and restoring ecosystems, and medical science devoted to improving human health and well-being, are prominent examples. Nonepistemic values concerning ethical goals seemingly permeate these kinds of sciences.

By way of illustration, one direct way in which ethical and sociopolitical values potentially bear on ecology (and possibly vice versa) is via population viability analyses (PVAs) (see Boyce 1992; Beissinger and McCullough 2002; Gerber and González-Suárez 2010). These are studies, usually model-based, of the dynamics of biological populations and how they would respond to various disturbance and management regimes. Endangered and threatened species are usual study targets. The northern spotted owl (*Strix occidentalis caurina*), for example, was and continues to be one of the most important and contentious test cases for the modeling approach. PVAs were integral in demonstrating the deleterious impact that continued logging of old-growth forests in the Pacific Northwest would have on the species (Lande 1988). These studies countered the official positions of various governmental agencies such as the US Forest Service, and eventually resulted in a court-ordered ban on logging.

Taking a step back, whether the available data are sufficient to show that a regime would adequately ensure a stipulated viability threshold is met usually requires a trade-off between minimizing type I and type II errors.[2] This in turn arguably requires the input of nonepistemic, ethical values. What else but the moral weight of making either kind of error – weightings that would vary considerably across contexts – could mark the appropriate trade-off? For instance, different degrees of vulnerability to threats, or different aesthetic, ecological, and economic properties of species, might merit different error trade-off thresholds. The scientific contribution PVAs make to conservation planning and action therefore seems to essentially incorporate ethical assumptions and considerations.

This is, of course, one example among many.[3] This chapter explores the range of ways values influence both ecological science and policy making informed by that science. Section 1 describes different ways in which nonepistemic values are claimed to influence science, and the difference between

[2] A type I error is rejecting a true "null" hypothesis (cases of so-called false positives). A type II error is failing to reject a false null hypothesis (cases of so-called false negatives).

[3] See Jamieson (2014) for an engaging and expansive analysis of similarly ethically fraught issues in climate science.

descriptive and ethically driven science. The previous chapter supplies the background for another example of such influence: how ethical values may determine the type 1/type II error trade-off in the application of island biogeography theory to the design of nature reserves (Shrader-Frechette 1990). Besides these trade-offs, choices of scientific categories and terms, for example, "carcinogen" (see Douglas 2000), "endangered," "invasive species" (see Boltovskoy et al. 2018), and even "pollutant" (Elliott 2011) are similarly infused with ethics. Numerous other examples could be cited, and philosophical interest in the role nonepistemic values should play in science, and actually do play, has recently transitioned from steady to burgeoning (see Proctor 1991; Shrader-Frechette 1991; Machamer and Wolters 2004; Koertge 2005; Kincaid et al. 2007; Carrier et al. 2008; Douglas 2009; Elliott 2011, 2017; Elliott and Richards 2017).

The clear impact ethical values can have has encouraged the view that ethically driven applied sciences are value-laden in an even stronger sense. It has been suggested that (1) both ethical values and nonnormative facts contribute *indispensably* to these sciences, so much so that (2) their respective contributions *cannot be demarcated*. In fact, the inextricable suffusion of value supposedly challenges a clear fact/value distinction. Some have also argued that this influence begets an unacceptable relativism in scientific testing in applied ecology: which hypotheses are ultimately accepted or rejected will be determined by the ethical evaluation of the relevant states of affair, such as whether species conservation is worth doing.

Sections 2 and 3 describe these charges and argue that they are overstated. In particular, they respond to two arguments that the role of values in ethically driven sciences shows that there is no defensible fact/value distinction. The first focuses on how uncertainty is managed in science, and the second appeals to a notion of fact–value "entanglement." The responses, it is hoped, (1) clarify the nature of the influence of nonepistemic values in ethically driven sciences and (2) show that this influence is no threat to a fact/value distinction. The value-laden character of ethically driven applied sciences also does not challenge the objectivity of hypothesis testing in them. Rather, although ethical values influence the general structure and methodologies of applied ecology, these influences can be demarcated from the factual status of claims made within it.

It is worth stressing that the view being defended here is arguably ubiquitous across scientific contexts. For example, it seems to be the rationale underlying this kind of guidance from the US Office of Management and Budget about peer review of policy documents:

> [W]here appropriate, reviewers should be asked to provide advice on the reasonableness of judgments made from the scientific evidence. However, the charge should make clear that the reviewers are not to provide advice on the policy (e.g. the amount of uncertainty that is acceptable or the amount of precaution that should be embedded in an analysis). Such considerations are the purview of the government. (quoted in Douglas 2005, 17)

The passage seems to presuppose the distinction at issue: between scientific inference of an ethically value-neutral kind and ethically value-laden considerations, in this case the acceptability of degrees of uncertainty and precaution that *should* occur. This kind of distinction is found throughout science. To cite just one more example, it seems to underlie the distinction between risk assessment and risk management often made within decision sciences: risk assessment being the scientific determination of what risks there are and their severity, and risk management being the formulation of what should be done about these risks, which obviously involves ethical values. For instance, in the early 1980s the US National Academy of Science (NAS) recommended that "regulatory agencies take steps to establish and maintain a clear conceptual distinction between assessment of risks and consideration of risk management alternatives; that is, the scientific findings and policy judgments embodied in risk assessments should be explicitly distinguished from the political, economic, and technical considerations that influence the design and choice of regulatory strategies" (quoted in Mayo 1991, 251).

This recommendation was in a report requested by the US Congress following a scandal in which President Reagan's leadership at the Environmental Protection Agency (EPA) was found to have perverted the scientific analysis of the carcinogenic risk of formaldehyde in favor of the chemical industry's interests (under the rubric of "more rigorous science," of course). Without a clear distinction between the scientifically factual and the ethically evaluative, the NAS worried there would be no basis to the criticism that the EPA distorted the *science*. The view defended in this chapter supports this kind of criticism by defending the fact/value distinction underlying it against recent criticisms from philosophers of science.

## 1 Descriptive and Ethically Driven Sciences

Before describing the numerous ways values shape sciences, ethically driven and not, great care is needed to delimit precisely what the controversial issue is. The topic of values in science is a minefield of ambiguities, confusions, and obscurities that demands cautious navigation. First, the principal concern is with the claimed role of *nonepistemic* values in science, not the functions of epistemic values such as empirical adequacy, explanatory power, generality, parsimony, theoretical unity, and the like. Epistemic values seem ubiquitous and indispensable in science, but this claim will not be addressed here.

Second, although many proponents of a fact/value distinction have been motivated by the view that there are no moral facts, whether via noncognitivism (Gibbard 1992) or some kind of error theory (Mackie 1977), the view defended in this chapter is neutral on that issue. So, for cognitivists who hold that moral claims are true or false and who are committed to the existence of moral facts, this analysis can be construed as attempting to establish a distinction between these normative facts, whatever their ultimate metaphysical status, and the type of descriptive facts assessed in science. The controversy is about the role of nonepistemic values in science, and thus the nature of science itself, not a metaethical dispute about whether irreducibly ethical facts and properties exist.

Third, science is not merely the activity of scientists. Scientists engage in a variety of activities: they testify to Congress, write recommendation letters, develop complex neuroses about their research, eat and excrete, and, unfortunately, sometimes also fabricate data and falsify records. These activities are unequivocally not part of science itself. That they legitimately or illegitimately involve nonepistemic values therefore does not threaten a clear fact/value distinction. What would seem to be a threat and will be our focus is if such values are essential in core aspects of science, such as in assessments of evidence and acceptance or rejection of hypotheses. Any contention that science upholds or challenges a distinction between descriptive facts and nonepistemic (ethical) values must keep the distinction between scientific activity and the mere activity of scientists in mind.

Similarly, apart from its compatibility with the view this chapter defends, the ethical and/or political issue of whether scientists *should* embrace nonepistemic values and integrate them more thoroughly into their work is not addressed. Several recent publications have argued, for instance, that scientists should consider the nonepistemic consequences of error in their work (see citations above). That may be right, but if so it would be the ethical conclusion of an ethical analysis (informed by the details of different scientific practices of course). It wouldn't seem to stem from an accurate appraisal of what scientists do *as scientists*. Scientists are human beings, and citizens, and community members, and employees, but they are also consumers of coffee, takers of antibiotics, and sometimes plagiarists. Just as it is clear the activities in the latter group indicate nothing about the practice of science, it is as unclear that the former group offers anything much more informative.

The ethical imperative to more thoroughly consider nonepistemic values in scientific work can be granted, however, without agreeing with the stronger, seemingly modal claims often accompanying this argument. For example, that "[evidence and nonepistemic values] are *inextricably* intermixed in accounts of the world" (Douglas 2007, 126) (emphasis added) or that "there is *no possibility* of drawing a sharp fact–value distinction [in science]" (Dupré 2007, 31) (emphasis added). Hilary Putnam (2002) has defended this view explicitly, labeling it "fact–value entanglement." As the potential allusion to quantum entanglement suggests, the idea is that descriptive facts and nonepistemic values cannot be clearly distinguished in science. This entanglement is made manifest, he and others believe, by appreciating the many influential ways such values shape applied sciences with ethical objectives. This chapter scrutinizes this claim by considering the discipline of applied ecology, and it is found wanting.

What is abundantly clear is that nonepistemic values play numerous roles within science. First and foremost, and like any other activity of *Homo sapiens*, what is considered ethically valuable, economically profitable, intellectually fascinating, or even just mildly stimulating shapes what science is about. Values therefore influence which phenomena, questions, hypotheses, or models are worth investigating. This influence of values and value-laden interests on what scientists do is uncontroversial because it is a precondition for their work, just as a desire to be financially compensated and, ideally, an interest in fixing intricate machinery accounts for why someone would pursue work as a mechanic. But recognizing how values catalyze attention and focus the efforts of these activities does not entail that there is an impact on the content of the work itself. Desires, interests, and values help explain why humans work as mechanics and scientists, but whether an adjustment fixes the engine trouble or a model's prediction is borne out depends on something else. The former role for values was accepted, for example, by paradigmatic advocates of a sharp and unbridgeable fact/value distinction such as Rudolf Carnap, Carl Hempel, and Ernest Nagel.[4]

The way in which values shape sciences is not monolithic; different values and interests beget different sciences with distinct characters. Two broad categories of science can be distinguished on this basis. What I will call *descriptive sciences* are principally concerned with discovering, describing, predicting, and explaining phenomena and regularities (DDPE). Canonical cases include physics, chemistry, biology, geology, paleontology, psychology, sociology, archaeology, and their assorted basic (as opposed to applied) subdisciplines, which for biology includes ecology, among many others.[5] The DDPE agenda obviously derives from our values, curiosity chief among them, and those same values targeted at different phenomena account for the different nature of descriptive sciences. Although manifested at starkly varying spatial and temporal scales, and ranging from structures deep within our brains to the farthest reaches of the universe, the main goal remains constant: better understanding the world around and within us.

Another class of sciences exhibits a different kind of teleology. What I will call *ethically driven sciences* are also concerned with discovering, describing, predicting, and explaining phenomena and regularities. But this agenda is pursued in the service of accomplishing goals considered ethically valuable. Examples include clinical psychology, conservation biology, numerous environmental sciences, invasion biology, medical sciences, restoration ecology, and welfare economics, among many others. The varying ethical objectives these sciences aspire to achieve largely account for their differing subject matters, priorities, and research methodologies. And while the goals are usually disparate, in some cases they can conflict. Restoring an ecosystem, for instance, requires fidelity to a historical benchmark (Desjardins 2015). The benchmark determines which species should be judged native, and what kinds of interventions are defensible. But a priority on native species and other historical specifics can mean restorative actions that are detrimental

[4] For an interesting discussion of nonepistemic value commitments and logical empiricism, see Roberts (2007).

[5] The distinction between basic and applied science can be problematic if not drawn carefully. As with any proposed classification, the distinction should be judged by the insights it affords and the research fruits it bears. See Niiniluoto (1993, 2013) for a fruitfully conceived distinction.

rather than beneficial to biodiversity conservation, the ultimate goal of conservation biology (see Chapters 5 and 6). Just as the goal shaping such a science is informed by ethical values, deciding how conflicting goals should be prioritized obviously requires ethical imput.

Since ethical values unsurprisingly play a formative and indispensable role in ethically driven sciences, certainly compared with descriptive sciences, it is important to consider just how pervasive and potent that role is. Specifically, does the explicitly ethical agenda of these sciences shed doubt on a fact/value distinction, especially given how significantly that agenda shapes their priorities and practices? Perhaps ethics serves as a universal acid, dissolving distinctions once deemed unassailable and tarnishing once lauded ideals, such as the value-free ideal (Kincaid et al. 2007) and the distinction between cognitive and noncognitive values itself (Longino 1996).

This general question spans several more specific ones concerning different aspects of scientific practice. For instance, do nonepistemic values in ethically driven sciences affect classification of phenomena such that their contribution to the task cannot be distinguished from what scientific realists would describe as the mind-independent properties of what is being studied? If yes, it seems the very categories scientists employ to investigate the world belie a fact/value distinction.

Statistical inference – the acceptance or rejection of hypotheses about phenomena in particular – is unquestionably at the core of scientific practice as well, and similar questions arise. It has been suggested, for example, that ethical values do affect hypothesis evaluation in a way that threatens a fact/value distinction by determining key features of statistical methodology, such as how uncertainty should be managed in Neyman–Pearson significance testing (e.g., the appropriate type I versus type II error threshold). Whether a given hypothesis is accepted or rejected could depend entirely on which ethical values are judged appropriate and actually applied; different values would yield different statistical inferences.

Most strikingly, besides hypothesis evaluation given uncertainty, perhaps nonepistemic values also influence the evidentiary basis for these statistical tests itself. On this view, values would legitimately affect assessments of truth in circumstances of complete certainty. Assessments of evidentiary strength and what constitutes evidence itself would be indelibly saturated with ethical value. The next two sections address these challenges to a fact/value distinction. Section 2 responds to the argument involving scientific inference. Section 3 responds to the even more formidable challenge of fact–value "entanglement," which would seem to establish the inseparable intertwining of fact and value in all the practices above: classification, statistical inference, and the very idea of evidence itself.

## 2 Uncertainty in Scientific Inference and "Inductive Risk"

The multifaceted debate described in the previous chapter was about more than the scientific support for the theory of island biogeography, its problematic application to reserve design, and overly narrow conceptions of what should count as progress in applied sciences with ethical objectives such as conservation biology. In a farsighted paper, Shrader-Frechette (1990) described how a disagreement over the proper role of values in science also contributed to the controversy. An analysis of extinction rates caused by deforestation in the tropics was the catalyst. Based on species-area curves and data on Costa Rican trees, Kangas (1986) argued that extinction rates were actually much lower than the estimates of other researchers. That, of course, struck several nerves. The ensuing exchanges trace many of the empirical and epistemic contours discussed in Chapter 6. The additional thread Shrader-Frechette put a finger on was a disagreement between Reed Noss and Dan Simberloff (among others) about whether minimizing type I or type II errors was the right approach when evaluating extinction hypotheses or possible reserve designs, especially within the backdrop of massive deforestation in the tropics.

Neither Noss nor Simberloff supplied much by way of philosophical argument for their views, according to Shrader-Frechette, so she helpfully outlined the stakes. Noss (1986), who accused Kangas of "bad science," thought type II errors should be minimized and type I errors should be risked in this ecologically, ethically, and politically charged context. If the null hypothesis were something like "deforestation at its current pace will not cause exorbitant extinction (say beyond some moderate expected rate X)" as it was for Kangas, then failing to reject it when it should be (i.e., it is false) would be a type I error.[6] Attempting to minimize type

[6] Note that besides null and alternate hypotheses being mutually exclusive, exhaustive, and other logical requirements, the framework of classical statistics itself provides no guidance about null hypothesis content, about which hypothesis should be considered the null. Different choices of null hypotheses can therefore result in dramatically different statistical judgments, and the absence of clear standards for selecting null

I rather than type II errors is therefore epistemically conservative. It prioritizes "asserting" less – in this statistical framework, rejecting fewer null hypotheses – in favor of greater confidence in the veracity of what is asserted. That outlook, Simberloff (1987) suggested, is more in keeping with scientific norms. Shrader-Frechette (1990) similarly judged it in accord with "scientific rationality."

Noss's criticism expresses a different and much broader kind of rationality. If the consequence of failing to reject a hypothesis is deleterious inaction in the face of serious threats – be it species extinctions or something else that impacts human welfare – he believed the priority should not be epistemic caution, but rather mitigating threats and achieving better outcomes. Shrader-Frechette labeled this "decision-theoretic rationality," the contrast being that the ethical status of consequences and their utility should counterbalance an exclusive focus on probabilistic assessment of hypotheses. The result may be that more false assertions are ultimately made, but gains in ethical consequences and utility enhancement outweigh the epistemic cost. As Shrader-Frechette (1990, 453) put it, "when one moves from pure science to applied science affecting policy, what is rational moves from epistemological considerations to both ethical and epistemological concerns."

This view has a long and important history, extant but underappreciated until the recent surge in interest in values in science. One of its first champions was Richard Rudner (1953), a student of Nelson Goodman, in a seminal paper "The Scientist qua Scientist Makes Value Judgments."[7] The basic case can be stated succinctly:

> [S]ince no scientific hypothesis is ever completely verified, in accepting a hypothesis the scientist must make the decision that the evidence is *sufficiently* strong or that the probability is *sufficiently* high to warrant the acceptance of the hypothesis. Obviously our decision ... is going to be a function of the importance, in the typically ethical sense, of making a mistake in accepting or rejecting the hypothesis. (1953, 2)

Douglas, Dupré, Elliott, Shrader-Frechette (in 1990 and more recently), and many others have expressed a similar view of the value-laden nature

hypotheses, or constructing null models that mathematically specify those hypotheses, has engendered much confusion. See Bausman (2018) for a useful discussion specific to ecology.

[7] The statistician C. West Churchman (1948, 1956) defended a similar position.

of scientific inference. Such examples are often referred to as cases of "inductive risk."

The way the most common statistical inference method used in science, Neyman–Pearson significance testing, manages uncertainty might seem to support this view. In Neyman–Pearson testing, the weight one attributes to making type 1 errors – rejecting a true null hypothesis – versus making type 2 errors – accepting a false null – can be expressed by specifying different alpha levels. A null hypothesis is rejected only if the *p*-value derived from the study is less than $\alpha$; smaller, more strict alphas therefore make rejecting a null more difficult, and increase the confidence that it should be rejected when it is (i.e., the null is false). In ethically driven sciences, these levels typically differ depending on the ethical severity of the consequences of making different types of error.[8] This accounts for more stringent alpha levels than 0.05, sometimes much more stringent, being used in some fields of medical research and nuclear engineering, for instance. Just as it is largely agnostic about the content of null hypotheses, the Neyman–Pearson framework does not necessitate a particular alpha level. So, rhetorically asks a proponent of decision-theoretic rationality like Rudner, on what other basis besides ethical import or perhaps economic import could such a specification be made? For example, after reviewing the ethically salient consequences of the trade-off between type 1 and 2 errors in studies of dioxin carcinogenicity, Douglas (2000, 569) claims that "determining the balance clearly requires an ethical value judgment in the internal stages of a scientific study." Hypothesis acceptance or rejection is clearly one such internal stage. Ethical values therefore seem to contribute indispensably to quintessentially scientific activities, activities that provide one, if not *the*, established standard for ascertaining what should count as descriptive facts. At least to the extent scientific practice was supposed to ground it, a sharp fact/value distinction appears moribund.

There are several responses to this argument. One is to deny that science is properly in the business of "accepting" and "rejecting" hypotheses. This was the view of Rudolf Carnap, Hans Reichbach, and Richard Jeffrey (1956), in an article that responded directly to Rudner. For them, the proper

[8] It is worth emphasizing that they can also vary for nonethical reasons. The hyper-stringent 5 sigma ($5\sigma$) significance criterion used in the detection of the Higgs boson, which is roughly a million times more severe than the standard 0.05 alpha level, seems to show that epistemic values alone *can* determine different alpha levels (see Franklin 2013).

objective of science is assigning probabilities to hypotheses directly, rather than accepting or rejecting them. Putting some sympathies with this view aside, the ubiquitous use of Neyman–Pearson significance testing and other classical, frequentist statistical methods to accept or reject hypotheses in science tells strongly against it. In fact, a staunch prior commitment to a sharp fact/value distinction seemed to be the primary motivation for their view of science's correct aims, whereas *defending* such a distinction is the goal of this analysis.

Before this methodologically austere view is unjustly relegated to the philosophical scrap heap, however, its natural home within a Bayesian framework is worth recognizing. For Bayesians, the proper assessment of the bearing of evidence $e$ on a hypothesis $h$ is typically how the evidence changes its posterior probability $p(h|e)$ as determined by Bayes's theorem $[p(h|e) = p(e|h)p(h)/p(e)]$, and as measured by a difference $[p(h|e) - p(h)]$, ratio $[p(h|e)/p(h)]$, or some other metric (see Fitelson 1999, 2007; Sober 2008, chapter 1). This requires estimating the relevant probabilities of the theorem. As is well known, doing so raises various theoretical difficulties in some scientific circumstances, especially concerning the prior probabilities (see Earman 1992, chapters 4–6). But none of these difficulties, nor different interpretations of what the relevant probabilities are (see Mellor 2005), provides a rationale by which nonepistemic values legitimately determine the values of these probabilities. The notion of accepting or rejecting a hypothesis, and the nonepistemic values it is claimed to require, is foreign to and unnecessary within the Bayesian statistical framework. Thus, to the extent Bayesianism is superior to frequentism, and perhaps the best statistical inference methodology for science in general,[9] this challenge to a fact/value distinction is a dead end. Unsurprisingly, this likely accounts for the primary focus on classical statistics and Neyman–Pearson significance testing by those recently critical of the distinction.

Bracketing the debate between Bayesianism and frequentism, let's grant that hypothesis acceptance and rejection is a legitimate and central activity of science as it is currently practiced. The first question is whether it *requires* nonepistemic values as has been suggested. More than fifty years ago, Isaac

[9] And this is not merely an academic question pursued by philosophers of science. Ecologists, conservation biologists, and environmental scientists are having the debate among themselves (see Wade 2000; Ellison 2004; Clark 2005; McCarthy 2007).

Levi (1960, 1962) persuasively argued that it does not. Behaviorism was then in its pervasive heyday, and Levi's main goal was to show that hypothesis acceptance or rejection does not require a behaviorist interpretation, such as the one defended by Neyman and Pearson themselves. Instead, it can be understood in terms of belief. What matters in the present context is his explanation that one's alpha level can be determined by strictly epistemic considerations in the search for truth. In particular, alpha may reflect the degree of caution sought in replacing nonbelief by true belief (recall footnote 8 above). A 0.01 alpha reflects more concern with mistakenly believing false hypotheses and less concern with replacing nonbelief by belief than a 0.05 or 0.1 alpha. Thresholds for accepting or rejecting hypotheses *can* be determined on a purely epistemic basis according to Levi, so the scientist qua scientist need not make value judgments, at least not in any "typically ethical sense."

At this point an objection could be raised about how hypothesis acceptance or rejection in science is being represented. Too much emphasis is being placed on the technical specifics of Neyman–Pearson significance testing and not enough on the fact that accepting or rejecting a hypothesis is *a decision* made by a scientist or group thereof. As a decision, it should be modeled decision-theoretically like any other. Doing so involves determining the utilities of scientists and any other stakeholders concerning the consequences of accepting or rejecting a given hypothesis. Together with the relevant probabilities, these utilities determine the expected utilities of "accepting" (or "rejecting") a hypothesis, which would provide the basis for deciding to do so. And since the utilities obviously depend on ethical and other values, hypothesis acceptance or rejection is nonepistemically value dependent.

To reveal how misguided this idea is, let us flesh out a simplistic but illuminative example. Consider a world in which there are only two potential disease treatments ($T_1$ and $T_2$) for pancreatic cancer, and only two possible hypotheses ($H_1$ and $H_2$) about the disease's pathology. Given that there are four possible pathology–treatment combinations, there are four different probabilities of producing a cure ($P_1$–$P_4$). In turn, since side effects are dependent on how the treatments causally interact with the pathological physiology underlying the disease (or at least we'll assume so), there are similarly four possible adverse side effects ($E_{11}$, $E_{12}$, $E_{21}$, and $E_{22}$). Let C designate the utility of being cured. Now imagine a cognitively troubled patient endeavoring to use this information not to make a choice between pursuing either treatment option, but rather to "accept" and then perhaps believe either $H_1$ or $H_2$ (see table below). With these parameters specified, the appropriate expected utility maximizing balance between the probability of a treatment's utility enhancing success and the consequences of its adverse side effects for the well-being of patients can be calculated. This information would then favor "accepting" one hypothesis over the other; all the ingredients for a decision are there.

Hypothesis "acceptance" or "rejection" by expected utility

| | $H_1$ | $H_2$ |
|---|---|---|
| $T_1$ | $P_1C - E_{11}$ | $P_2C - E_{12}$ |
| $T_2$ | $P_3C - E_{21}$ | $P_4C - E_{22}$ |

As a characterization of *scientific* hypothesis acceptance or rejection, something has gone terribly wrong. In science at least, acceptance or rejection does not depend on the nonepistemic utilities associated with doing so such that the high utility of accepting a hypothesis can *trump* its low probability, or the high utility of rejecting a hypothesis can trump its high probability. Rather, it primarily depends on the likelihood the hypothesis being accepted or rejected is actually true or false. If the utilities are epistemically restricted in this way, that is, if the goal is *only* to replace nonbelief by true belief and false belief by nonbelief (or the more sophisticated probabilistic counterpart of this categorial objective [see Pettigrew 2016]), the utilities in question (suitably normalized) would be simple: 1 for believing $H_1$ or $H_2$ when either is true and 0 for believing either when either is false. The inference would then be solely determined by the probabilities of either hypothesis. The pathological case described above is so problematic because the nonepistemic factors eclipse the epistemic factors. Note also that a third option would always be available in such a scenario if it were correctly characterized: remaining agnostic. The utility assigned to remaining agnostic would depend only on the epistemic value trade-off between mistakenly believing false hypotheses versus replacing nonbelief by belief.

That nonepistemic values *need not* play a role in these methods of statistical inference, that strictly epistemic ones suffice and the decision can therefore be represented as a purely epistemic one, prompts some doubt that descriptive facts and nonepistemic values are *inextricable* in ethically driven sciences. But it is clear that in many cases nonepistemic values do influence the elements of hypothesis evaluation we have discussed, and do so legitimately. In such cases they do not outweigh epistemic elements, but rather contribute when uncertainty makes epistemic considerations insufficiently effectual. If a hypothesis says a treatment will have no remedial effect on pancreatic cancer, and mistakenly rejecting it results in ineffectively treated cancer patients and increased pain, this certainly justifies and is used to stipulate a more stringent alpha level. In such scenarios, nonepistemic values fill the decision-theoretic gaps epistemic values cannot span. Given that these ethically salient cases are found throughout applied ecology and the ethically driven sciences more generally, doesn't this establish inextricable fact–value entanglement?

The answer is no, but the nature of what counts as descriptively factual must be rethought. Since whether a hypothesis is accepted or rejected clearly depends on the input of nonepistemic values in these cases, particularly ethical ones, preserving a fact/value distinction requires showing that this dependency is not entanglement, that a descriptively factual and a nonepistemically evaluative component of this process can be identified. Where nonepistemic values (legitimately) influence statistical inference, severing the entanglement but preserving the dependency is achieved by a type of conditionalization.

Specifically, hypothesis acceptance or rejection resolves into several components. The proposal is that what remains descriptively factual is the following:

(1) the data available for evaluating a hypothesis (*D*)
(2) a probabilistic conditional claim $[(H \wedge S \wedge D) \rightarrow O]$ whose antecedent represents:
   (i) which hypothesis is being evaluated (*H*)
   (ii) the statistical test used to evaluate it (*S*)
   (iii) the data involved in the test (*D*),
   and whose consequent represents the test's outcome (*O*).

For lack of a better term, call these conditional claims "inference conditionals."[10] Given a hypothesis being assessed and a statistical test doing the

[10] Note that they can be probabilistic. Multiple iterations of some nondeterministic statistical tests, such as those employing Monte Carlo randomization methods, may yield different outcomes for the same hypothesis and data (see Robert and Casella 2004).

assessing notice also that items (1) and (2) (probabilistically) entail a test's outcome to accept or reject a hypothesis. The value component is the claim that the statistical test being performed given this hypothesis and data adequately reflects the nonepistemic values relevant to the hypothesis being tested, such as the ethical status of consequences of erroneous test outcomes.

Most of the philosophically heavy lifting is done by the inference conditional, but a remark about the first descriptive element is needed. One could object that the data available will not always be strictly descriptive. In some cases, data collection is limited by ethical restrictions, for instance, on the use of human and animal subjects. Whether one has these data versus other data is therefore influenced by nonepistemic values. This is uncontroversial, but the effect is on what data are available, not on what could be considered the data's content: what the data say about the phenomena being studied. On the other hand, in the way described earlier, nonepistemic values do seem to effect statistical testing in a stronger sense by affecting a test's outcomes – whether a hypothesis is rejected or accepted at all. There thus seems to be an important disanalogy between the limited influence of nonepistemic values on what data are available and their influence on statistical testing of hypotheses, and thus no threat to the descriptive status of the available data. There is a potential complication here concerning the concepts involved in acquiring and interpreting data that the next section addresses.

Turning to inference conditionals, what grounds the claim that they are descriptively factual, unentangled by nonepistemic values? The case rests on the features these conditionals share with statements commonly and uncontroversially considered descriptively factual. First, these conditionals provide an intersubjective basis for agreement that is typically possessed by descriptive claims but often absent concerning ethical ones. Imagine two scientists disagreeing whether some treatment for pancreatic cancer *should* be accepted as effective, or whether habitat loss and concurrent factors or something else was responsible for an endangered species' declining population size (e.g., Brown et al. 2017). Both used, say, Neyman–Pearson significance testing on the same well-vetted data, but one used a 0.05 alpha and rejected the null that the treatment would produce no difference in cancer growth or that the pollutant was causally inert for the species, while the other used a 0.01 alpha and failed to reject. Since they used the same data, evaluated the same hypothesis, and obviously agree they performed different statistical tests, the disagreement must be located in the relevant inference conditional or the claim that the test used reflects the appropriate ethical considerations, the "value element" as I have labeled it. But to locate the disagreement in the conditionals would be to deny what each scientist in fact did, which they both assent to, and which is a precondition of their disagreement. Rather, the disagreement most plausibly stems from different ethical reasons for performing one statistical test over another (specifically, using different alphas). To the extent that intersubjective convergence tracks the descriptively factual and divergence is indicative of the ethically evaluative – contrast that marks *the* fundamental difference between the two for someone like Bernard Williams (1986), for example – this supports the distinction drawn. Note that, following Williams, this argument does not presuppose that intersubjective convergence *cannot* be reached on ethical questions in general, or that it *must* on descriptive matters.

Second, the truth value of the conditionals seems to be transparently empirically scrutable in the same way as other uncontroversially descriptive claims in science. The phrase "transparently empirically scrutable" requires some elaboration. By way of example, the claim that the mean height of the humans in some classroom falls in the interval from 5′3″ to 5′10″ is transparently scrutable. Each step in the process of determining that mean is transparently ascertainable using empirical methods. Each human's height can be estimated with a tape measure. Those measurements can be recorded in a ledger, and then the mean can be calculated using the standard mathematical equation for the mean. The process can then be repeated if necessary. In terms of transparent empirical scrutiny, the inference conditionals are no different. With the data, hypothesis, and the specifics of a statistical test given, implementing the test with respect to the data and hypothesis is just as straightforwardly calculational as using the formula for the mean. It is, of course, computationally much more intensive and usually executed with software. But the software's code executing the test and the internal processes that eventually result in an output can be inspected and scrutinized.[11] Statistics and arithmetic are no different in kind on the metric of empirical transparency.

In contrast, whether or not ethical claims can be true or false, their truth value does not seem to be similarly transparently scrutable. In the present

[11] Calculating the $p$-value could also be done manually, although that would be far less reliable than using software.

context, that a particular statistical test appropriately reflects the ethical weights that should be placed on different types of potential errors in rejecting or accepting hypotheses is not empirically or even conceptually tractable in the same transparent way. This is particularly apparent if, as critics of a fact/value distinction such as Hilary Putnam often emphasize and the next section discusses, grasping ethical concepts requires acquiring and being able to have certain experiences within an evaluative "first person" perspective.

These considerations support a distinction between the descriptively factual and the nonepistemically valuative for which traditional supporters of the distinction might have little affinity. As the distinction has been standardly portrayed, caricatured really, it appeals to an incontrovertible criterion for descriptivity, such as one focused on so-called elementary experiences or purely observational statements as supposedly utilized by some of the early logical empiricists (see Oberdan 1990). Only with such a criterion can a definitive separation be made between facts, which science is concerned with, and values. A similarly incontrovertible criterion for ethical values, perhaps that they are intrinsically motivating, would also buttress such a sharp division. But these ambitious criteria have proven unworkable. The expectation that facts and values can only be defensibly distinguished on such absolutist grounds, however, should be resisted.

## 3 Scientific Concepts and Fact–Value "Entanglement"

Even if the argument in Section 2 is compelling, it might be argued there is a stronger and logically prior case for fact–value entanglement at the level of concepts, one forcefully advanced by Hilary Putnam (e.g., 2002) and endorsed by John Dupré (2007), among others. The argument proceeds as follows. First, hypotheses, models, and theories are formulated with concepts, and concepts shape the acquisition and interpretation of data as well. But, second, for some "thick" concepts (Eklund 2011) the descriptive and ethically evaluative are indelibly intertwined. Concepts within some ethically driven human sciences, such as *health* in medicine (Murphy 2015), *well-being* in welfare economics and other disciplines (Adler and Fleurbaey 2016), and *abuse* and *disorder* as analyzed in clinical psychology (Wakefield 1992), are frequently cited as examples. The ethically suffused complexity of the concept of *biodiversity* makes it a likely candidate (see Chapter 5). Given the conceptual confusion and argumentative heat it has generated, perhaps the concept of an *invasive species* in invasion biology also merits the status (Boltovskoy et al. 2018). Even some concepts in descriptive sciences seem to share this intertwining character, such as in studies of forced mating in the animal kingdom. In these cases, the supposedly descriptive elements from above (the data available, the hypothesis, and the inference conditional) are suffused with ethical values, so the proposed distinction between descriptive facts and nonepistemic values fails.

The first premise is undeniable. The second is not. Hilary Putnam has been its most prominent exponent. Although Putnam's primary concern was criticizing the fact/value distinction as conceptualized by noncognitivists – that is, as a distinction between objective facts and merely subjective, emotive attitudes or expressions – the argument is also intended to apply to the kind of distinction this chapter defends. The problem is with so-called thick ethical concepts:

> What is characteristic of "negative" descriptions like "cruel," as well as of "positive" descriptions like "brave," "temperate," and "just" ... is that to use them with any discrimination one has to be able to identify imaginatively with an *evaluative point of view*. (Putnam 2002, 39; emphasis added)

Putnam goes on to argue that these concepts permeate scientific disciplines, welfare economics being his particular case study. Dupré (2007) has echoed this judgment and cited proposed explanations of rape in evolutionary psychology as a clear case.

To assess this argument, the first question that needs answering is exactly what "identify[ing] imaginatively with an evaluative point of view" requires. To represent the behavior of an orangutan accurately, for instance, a primatologist probably needs to creatively think like one, imaginatively conjecturing various social alliances, threats, food resources, and so on. But presumably this exercise in imagination would be no threat to a clear fact/value distinction. Putnam's (2002) book does not elaborate the charge, but an earlier essay provides a bit of helpful detail:

> [W]hen we are actually confronted with situations requiring ethical evaluation ... the sorts of descriptions that we need – descriptions of the motives and character of human beings, above all – are descriptions in the language of a "sensitive novelist," not in scientistic or bureaucratic jargon. (Putnam 1990, 166)

The idea then seems to be that "scientistic" descriptions, which are presumably ones that typically attempt to maximize things like precision and measurability, and that are characteristically formulated from a third-person perspective, cannot adequately represent thick concepts, which require, the passage seems to suggest, the first-person perspective characteristic of a "sensitive novelist."

With this understanding of the criticism, however, it is unclear what difficulty is posed for the distinction defended above. The issue can be framed in the following way. If the "imaginative identification with an evaluative point of view" requires a first-person perspective as it seems to for Putnam, then there seem to be two possibilities. The first is that the kind of first-person perspective required is the one often found in discussions of philosophy of mind or moral phenomenology and commonly taken to be *in principle* irreducible to, and unanalyzable from, the third-person perspective available in science (see Nagel 1974; Chalmers 1997). If so, then it is commonly thought that science is at a loss, necessarily ill-equipped to investigate these concepts.[12] Perhaps poor, inadequate surrogates can be formulated and analyzed from the third-person perspective, but such efforts would be judged unlikely to yield any valuable insights given the fundamental obstacle the first-person perspective would present. In this case, it is not that fact and value collapse; there is simply collapse from the point of view of what science can achieve.

If, on the other hand, the first-person perspective required is not scientifically unanalyzable, then the distinction from above, modified slightly, seems to account for how this could occur. The modification required simply makes two additions to the value side of the distinction: that (1) the concepts used in the formulation of the hypothesis and (2) those shaping the acquisition and interpretation of the data reflect the relevant nonepistemic values. For a given "thick" concept, "well-being," for example, those with the requisite expertise in the relevant first-person perspective, professional ethicists or novelists perhaps, could assist scientists in selecting which third-person-perspective surrogate – usually an empirically measurable or otherwise

[12] Although this pessimistic appraisal neared orthodoxy several decades ago, among those immersed in the relevant scientific fields it no longer appears nearly as plausible. For a scientifically rich, penetrating, and intellectually stimulating counterpoint, one that challenges the cogency of the first-person/third-person distinction itself, see Peter Godfrey-Smith (2013, 2016, 2019).

tractable quantity – most accurately captures the concept's content, assuming it can be captured. Scientists would have an important role assisting in this selection, by illustrating and explaining the properties of various measures, for instance. If a set of such surrogates is equally adequate, the consequences of using each on test outcomes can be determined. That there may not be a unique best surrogate within the third-person perspective, or perhaps even no adequate surrogate at all, would reveal a serious limitation of scientific analyses of such concepts, but would not challenge the descriptive fact/nonepistemic value distinction.

# References


Abele, L. and Connor, E. F. (1979), "Application of Island Biogeography Theory to Refuge Design: Making the Right Decision for the Wrong Reasons." In R. M. Linn (ed.), *Proceedings of the First Conference on Scientific Research in National Parks.* U.S.D.I. National Park Service Transactions and Proceedings. No. 5: 89–94.

Abler, R. (1987), "What Shall We Say? To Whom Shall We Speak?" *Annals of the American Association of Geographers* **77**: 511–524.

Abrams, P. (1983), "The Theory of Limiting Similarity." *Annual Review of Ecology and Systematics* **14**: 359–376.

Ackery, P. R. and Vane-Wright, R. I. (1984), *Milkweed Butterflies*. Ithaca, NY: Cornell University Press.

Adler, M. and Fleurbaey, M. (eds.), (2016), *Oxford Handbook of Wellbeing and Public Policy*. New York: Oxford University Press.

Alexander, J. and Skyrms, B. (1999), "Bargaining with Neighbors: Is Justice Contagious?" *Journal of Philosophy* **96**: 588–598.

Ankeny, R. (2001), "Model Organisms as Cases: Understanding the Lingua Franca at the Heart of the Human Genome Project." *Philosophy of Science* 68(3 Suppl.), s251–s261.

Ankeny, R. and Leonelli, S. (2011), "What's so Special about Model Organisms?" *Studies in History and Philosophy of Science* **42**: 313–323.

Arthur, W. (1990), *The Green Machine: Ecology and the Balance of Nature*. Cambridge, MA: Basil Blackwell.

Auerbach, S. I. (1972), "Ecology, Ecologists, and the ESA." *Ecology* **53**: 205–207.

Azzouni, J. (2004), *Deflating Existential Consequence*. New York: Oxford University Press.

Bausman, W. (2018), "Modeling: Neutral, Null, and Baseline." *Philosophy of Science* 85: 594–616.

Beatty, J. (1995), "The Evolutionary Contingency Thesis." In G. Wolters and J. Lennox (eds.), *Concepts, Theories, and Rationality in the Biological Sciences*. Pittsburgh, PA: University of Pittsburgh Press, pp. 49–81.

(2006), "Replaying Life's Tape." *Journal of Philosophy* **103**: 336–362.

Beissinger, S. R. and McCullough, D. R. (eds.) (2002), *Population Viability Analysis*. Chicago: University of Chicago Press.

Bensaude-Vincent, B. and Stengers, I. (1996), *A History of Chemistry*. Cambridge, MA: Harvard University Press.

Bird, A. (2007), "What Is Scientific Progress?" *Nous* **41**: 64–89.

Boltovskoy, D.; Sylvester, F.; and Paolucci, E. (2018), "Invasive Species Denialism: Sorting Out Facts, Beliefs, and Definitions." *Ecology and Evolution* **8**: 11190–11198.

Bormann, F. H. (1971), "President's Report." *Bulletin of the ESA* **52**: 1–4.

Bowler, P. (1976), "Malthus, Darwin, and the Concept of Struggle." *Journal of the History of Ideas* **37**: 631–650.

Boyce, M. (1992), "Population Viability Analysis." *Annual Review of Ecology and Systematics* **23**: 481–497.

Brown, D.; Ribic, C.; Donner, D.; Nelson, M.; Bocetti, C.; and Deloria-Sheffield, C. (2017), "Using a Full Annual Cycle Model to Evaluate Long-Term Population Viability of the Conservation-Reliant Kirtland's Warbler after Successful Recovery." *Journal of Applied Ecology* **54**: 439–449.

Brun, G. (2016), "Explication as a Method of Conceptual Re-engineering." *Erkenntnis* **81**: 1211–1241.

Budiansky, S. (1995), *Nature's Keepers: The New Science of Nature Management*. New York: Free Press.

Burley, F. W. (1988), "Monitoring Biological Diversity for Setting Conservation Priorities." In E. O. Wilson (ed.), *Biodiversity*. Washington, DC: National Academy Press, pp. 277–230.

Busby, J. R. (1991), "BIOCLIM – A Bioclimatic Analysis and Prediction System." In C. Margules and M. P. Austin (eds.), *Nature Conservation: Cost-Effective Biological Surveys and Data Analysis*. Chipping Norton: Surrey Beatty & Sons, pp. 64–67.

Carnap, R. (1962), *The Logical Foundations of Probability*. 2nd edition. Chicago: University of Chicago Press.

Carrier, M.; Howard, D.; and Kourany, J. (eds.) (2008), *The Challenge of the Social and the Pressure of Practice*. Pittsburgh, PA: University of Pittsburgh Press.

Caughley, G. (1994), "Directions in Conservation Biology." *The Journal of Animal Ecology* **63**: 215–244.

Chalmers, D. (1997), *The Conscious Mind: In Search of a Fundamental Theory*. New York: Oxford University Press.

Chandrasekharan, S. and Nersessian, N. (2015), "Building Cognition: The Construction of Computational Representations for Scientific Discovery." *Cognitive Science* **39**: 1727–1763.

Chapin, F., III; Matson, P.; and Vitousek, P. (2011), *Principles of Terrestrial Ecosystem Ecology*. New York: Springer.

Chapman, R. and Wylie, A. (2016), *Evidential Reasoning in Archeology*. New York: Bloomsbury Academic.

Chase, J. and Leibold, M. (2003), *Ecological Niches: Linking Classical and Contemporary Approaches*. Chicago: University of Chicago Press.

Churchman, C. W. (1948), *Theory of Experimental Inference*. New York: Macmillan.

(1956), "Science and Decision Making." *Philosophy of Science* **23**: 247–249.

Clark, J. (2005), "Why Environmental Scientists Are Becoming Bayesians." *Ecology Letters* **8**: 2–14.

Clements, F. E. (1916), *Plant Succession: An Analysis of the Development of Vegetation*. Washington, DC: Carnegie Institute of Washington.

Cocks, K. D. and Baird, I. A. (1989), "Using Mathematical Programming to Address the Multiple Reserve Selection Problem: An Example from the Eyre Peninsula, South Australia." *Biological Conservation* **49**: 113–130.

Cocks, K. D. and Ive, J. R. (1988), "Evaluating a Computer Package for Planning Public Lands in New South Wales." *Journal of Environmental Management* **26**: 249–260.

Cocks, K. D.; Ive, J. R.; Davis, J. R.; and Baird, I. A. (1983), "SIRO-PLAN and LUPLAN: An Australian Approach to Land Use Planning. 1. The SIRO-PLAN Land Use Planning Method." *Environment and Planning B* **10**: 331–345.

Colinvaux, P. (1979), *Why Big Fierce Animals Are Rare: An Ecologist's Perspective*. Princeton: Princeton University Press.

Colwell, R. (1992), "Niche: A Bifurcation in the Conceptual Lineage of the Term." In E. F. Keller and E. S. Lloyd (eds.), *Key Words in Evolutionary Biology*. Cambridge, MA: Harvard University Press, pp. 241–248.

Connor, E. and McCoy, E. (1979), "The Statistics and Biology of the Species–Area Relationship." *American Naturalist* **113**: 791–833.

Cooper, G. (2003), *The Science of the Struggle for Existence: On the Foundations of Ecology*. New York: Cambridge University Press.

Coyne, J. and Allen, H. (2004), *Speciation*. Sunderland: Sinauer Associates.

Crowcroft, P. (1991), *Elton's Ecologists: A History of the Bureau of Animal Population*. Chicago: University of Chicago Press.

Csuti, B. (1994), "Gap Analysis: Mapping Biodiversity for Conservation and Management." *Endangered Species Update* **11**: 1–4.

Cummins, R. (1975), "Functional Analysis." *Journal of Philosophy* **75**: 741–765.

(1996), *Representations, Targets, and Attitudes*. Cambridge, MA: MIT Press.

Currie, A. (2018), *Rock, Bone, and Ruin: An Optimist's Guide to the Historical Sciences*. Cambridge, MA: MIT Press.

Darwin, C. (1859), *On the Origin of the Species*. Cambridge, MA: Harvard University Press.

Davis, F. W.; Davis, D. M.; Stoms, J. E.; Estes, J. E.; Scepan, J.; and Scott, J. M. (1990), "An Information Systems Approach to the Preservation of Biological Diversity." *International Journal of Geographical Information Systems* **4**: 55–78.

Davis, M. and Shaw, R. (2001), "Range Shifts and Adaptive Responses to Quaternary Climate Change." *Science* **292**: 673–679.

DeAngelis, D. and Waterhouse, J. (1987), "Equilibrium and Non-equilibrium Concepts in Ecological Models." *Ecological Monographs* **57**: 1–21.

Dejardins, E. (2015), "Historicity and Ecological Restoration." *Biology and Philosophy* **30**: 77–98.

Diamond, J. (1975), "The Island Dilemma: Lessons of Modern Biogeographic Studies for the Design of Natural Reserves." *Biological Conservation* **7**: 129–146.

(1976), "Island Biogeography and Conservation: Strategy and Limitations." *Science* **193**: 1027–1029.

Diamond, J. and May, R. (1976), "Island Biogeography and the Design of Natural Reserves." In R. May (ed.), *Theoretical Ecology: Principles and Applications*. New York: Blackwell, pp. 163–186.

Dietrich, M. (1998), "Paradox and Persuasion: Negotiating the Place of Molecular Evolution within Evolutionary Biology." *Journal of the History of Biology* **31**: 85–111.

Dijksterhuis, E. (1961), *The Mechanization of the World Picture*. New York: Oxford University Press.

Dobzhansky, T. (1964), "Biology, Molecular and Organismic." *American Zoologist* **4**: 443–452.

(1973), "Nothing in Biology Makes Sense Except in the Light of Evolution." *American Biology Teacher* **35**: 125–129.

Dodds, P.; Rothman, D.; and Weitz, J. (2001), "Re-examination of the '3/4-law' of Metabolism." *Journal of Theoretical Biology* **209**: 9–27.

Douglas, H. (2000), "Inductive Risk and Values in Science." *Philosophy of Science* **67**: 559–579.

(2005), "Boundaries between Science and Policy: Descriptive Difficulty and Normative Desirability." *Environmental Philosophy* **2**: 14–29.

(2007), "Rejecting the Ideal of Value-Free Science." In H. Kincaid, J. Dupré, and A. Wylie (eds.), *Value-Free Science? Ideals and Illusions*. New York: Oxford University Press, pp. 120–139.

(2009), *Science, Policy, and the Value-Free Ideal*. Pittsburgh, PA: University of Pittsburgh Press.

Downes, S. and Matthews, L. (2019), "Heritability." In *The Stanford Encyclopedia of Philosophy* (Winter 2019 edition), E. N. Zalta (ed.), https://plato.stanford.edu/archives/win2019/entries/heredity/.

Drake, J.; Fuller, M.; Zimmerman, C.; and Gamarra, J. (2007), "Emergence in Ecological Systems." In N. Rooney, K. McCann, and D. Noakes (eds.), *From Energetics to Ecosystems: The Dynamics and Structure of Ecological Systems*. New York: Springer, pp. 157–184.

Dritschilo, W. (2008), "Bringing Statistical Methods to Community and Evolutionary Ecology: Daniel S. Simberloff." In O. Harman and M. R. Dietrich (eds.), *Rebels, Mavericks, and Heretics in Biology*. New Haven, CT: Yale University Press, pp. 356–371.

Dupré, J. (2007), "Fact and Value." In H. Kincaid, J. Dupré, and A. Wylie (eds.), *Value-Free Science? Ideals and Illusions*. New York: Oxford University Press, pp. 27–41.

Dussault, A. (2018), "Functional Ecology's Non-selectionist Understanding of Function." *Studies in History and Philosophy of Science Part C* **70**: 1–9.

Dyer, L. A. and Coley, P. D. (2001), "Latitudinal Gradients in Tri-trophic Interactions." In T. Tscharntke and B. A. Hawkins (eds.), *Multitrophic Level Interactions*. New York: Cambridge University Press, pp. 67–88.

Earman, J. (1986), *A Primer on Determinism*. New York: Springer.

(1992), *Bayes or Bust? A Critical Examination of Bayesian Confirmation Theory*. Cambridge, MA: MIT Press.

Earman, J.; Glymour, C.; and Mitchell, S. (eds.), (2002), *Ceteris Paribus Laws*. Dordrecht: Kluwer.

Eberhardt, L. L. (1969), "Some Aspects of Species Diversity Models." *Ecology* **50**: 503–505.

Egerton, F. (1973), "Changing Concepts of the Balance of Nature." *Quarterly Review of Biology* **48**: 322–350.

Ehrenfeld, D. (1976), "The Conservation of Non-Resources." *American Scientist* **64**: 648–656.

Ehrlich, P. and Ehrlich, A. (1981), *Extinction*. New York: Random House.

Eklund, M. (2011), "What Are Thick Concepts?" *Canadian Journal of Philosophy* **41**: 25–50.

Elgin, M. (2006), "There May Be Strict Empirical Laws in Biology, after All." *Biology and Philosophy* **21**: 119–134.

Eliot, C. (2011a), "The Legend of Order and Chaos: Communities and Early Community Ecology." In K. de Laplante, B. Brown, and K. Peacocke (eds.),

*Handbook of the Philosophy of Science, vol. 11: Philosophy of Ecology*. Amsterdam: Elsevier, pp. 49–108.

(2011b), "Competition Theory and Channeling Explanation." *Philosophy & Theory in Biology* **3**: 1–16.

Elliott, K. (2011), *Is a Little Pollution Good for You? Incorporating Societal Values in Environmental Research*. New York: Oxford University Press.

(2017), *A Tapestry of Values: An Introduction to Values in Science*. New York: Oxford University Press.

Elliott, K. and Richards, T. (eds.) (2017), *Exploring Inductive Risk: Case Studies of Values in Science*. New York: Oxford University Press.

Elliott-Graves, A. (2016), "The Problem of Prediction in Invasion Biology." *Biology and Philosophy* **31**: 373–393.

(2020), "The Value of Imprecise Prediction." *Philosophy, Theory, and Practice in Biology* **12**(4).

Ellison, A. (2004), "Bayesian Inference in Ecology." *Ecology Letters* **7**: 509–520.

Elton, C. ([1927] 2001), *Animal Ecology*. Chicago: University of Chicago Press.

(1954), "An Ecological Textbook." *Journal of Animal Ecology* **23**: 282–284.

Epstein, B. (2012), "Agent-Based Modeling and the Fallacies of Individualism." In P. Humphreys and C. Imbert (eds.), *Models, Simulations, and Representations*. New York: Routledge, pp. 115–144.

Ereshefsky, M. (2001), *The Poverty of the Linnaean Hierarchy*. Cambridge, MA: Cambridge University Press.

Ferrier, S. (1988), *Environmental Resource Mapping System (E-RMS), Users Manual for Version 1.2*. NSW National Parks and Wildlife Service.

(1990), "Applying GIS to Environmental Survey Design, Analysis and Extrapolation." *URPIS* **18**: 293–301.

(1992), *Development of a Predictive Modeling Module for E-RMS*. New South Wales National Parks and Wildlife Service Report to the Australian National Parks and Wildlife Service.

Ferrier, S. and Smith, A. (1990), "Using Geographical Information Systems for Biological Survey Design, Analysis and Extrapolation." *Australian Biologist* **3**: 105–116.

Finkel, E. (1998a), "Software Helps Australia Manage Forest Debate." *Science* **281**: 1789–1791.

(1998b), "Forest Pact Bypasses Computer Model." *Science* **282**: 1968–1969.

Fitelson, B. (1999), "The Plurality of Bayesian Measures of Confirmation and the Problem of Measure Sensitivity." *Philosophy of Science* (Proceedings Supplement) **66**: S362–S378.

(2007), "Likelihoodism, Bayesianism, and Relational Confirmation." *Synthese* **156**: 473–489.

Franklin, A. (2013), *Shifting Standards: Experiments in Particle Physics in the Twentieth Century*. Pittsburgh, PA: University of Pittsburgh Press.

Fretwell, S. F. (1987), "Food Chain Dynamics: The Central Theory of Ecology?" *Oikos* **50**: 291–301.

Gause, G. ([1934] 2019), *The Struggle for Existence*. Mineola, NY: Dover.

Gerber, L. and González-Suárez, M. (2010), "Population Viability Analysis: Origins and Contributions." *Nature Education Knowledge* **3**: 15.

Gibbard, A. (1992), *Wise Choices, Apt Feelings: A Theory of Normative Judgment*. Cambridge, MA: Harvard University Press.

Gilbert, F. S. (1980), "The Equilibrium Theory of Island Biogeography: Fact or Fiction?" *Journal of Biogeography* **7**: 209–235.

Gilpin, M. E. and Diamond, J. (1980), "Subdivision of Nature Reserves and the Maintenance of Species Diversity." *Nature* **285**: 567–568.

Ginzburg, L. R. (1972), "The Analogies of the 'Free Motion' and 'Force' Concepts in Population Theory" (in Russian). In V. A. Ratner (ed.), *Studies on Theoretical Genetics*. Novosibirsk: Academy of Sciences of the USSR, pp. 65–85.

(1986), "The Theory of Population Dynamics: I. Back to First Principles." *Journal of Theoretical Biology* **122**: 385–399.

Ginzburg, L. R. and Colyvan, M. (2004), *Ecological Orbits: How Planets Move and Populations Grow*. New York: Oxford University Press.

Glazier, D. (2006), "The 3/4-Power Law Is Not Universal: Evolution of Isometric, Ontogenetic Metabolic Scaling in Pelagic Animals." *BioScience* **56**: 325–332.

Godfrey-Smith, P. (2006), "The Strategy of Model-Based Science." *Biology and Philosophy* **21**: 725–740.

(2007), "Conditions for Evolution by Natural Selection." *Journal of Philosophy* **104**: 489–516.

(2013), "On Being an Octopus: Diving Deep in Search of the Human Mind." *Boston Review* (June 3), http://bostonreview.net/books-ideas/peter-godfrey-smith-being-octopus.

(2016), *Other Minds: The Octopus, the Sea, and the Deep Origins of Consciousness*. New York: Farrar, Straus and Giroux.

(2019), "Evolving across the Explanatory Gap." *Philosophy, Theory, and Practice in Biology* **11**: 1.

Goh, B. S. (1977), "Global Stability in Many-Species Systems." *American Naturalist* **111**: 135–143.

Gómez-Pompa, A.; Vázquez-Yanes, C.; and Guevera, S. (1972), "The Tropical Rain Forest: A Nonrenewable Resource." *Science* **177**: 762–765.

Griesemer, J. (1990), "Material Models in Biology." *PSA Proceedings* **2**: 79–93.

(1992), "Niche: Historical Perspectives." In E. F. Keller and E. S. Lloyd (eds.), *Key Words in Evolutionary Biology*. Cambridge, MA: Harvard University Press, pp. 231–240.

Grimm, V. and Railsback, S. (2005), *Individual-Based Modeling and Ecology*. Princeton: Princeton University Press.

Grinnell, J. (1904), "The Origin and Distribution of the Chestnut-Backed Chickadee." *The Auk* **21**: 375–377.

(1917), "The Niche-Relationships of the California Thrasher." *The Auk* **34**: 427–433.

(1924), "Geography and Evolution." *Ecology* **5**: 225–229.

Gupta, A. (2015), "Definitions." In *The Stanford Encyclopedia of Philosophy* (Winter 2019 edition), E. N. Zalta (ed.), https://plato.stanford.edu/archives/win2019/entries/definitions/.

Haeckel, E. (1866), *General Morphology of Organisms*. Berlin.

Hacking, I. (1979), "What Is Logic?" *Journal of Philosophy* **76**: 285–319.

Hagen, J. (1989), "Research Perspectives and the Anomalous Status of Modern Ecology." *Biology and Philosophy* **4**: 433–455.

(1992), *An Entangled Bank: The Origins of Ecosystem Ecology*. Newark, NJ: Rutgers University Press.

Haldane, J. B. S. (1924), "A Mathematical Theory of Natural and Artificial Selection: Part I." *Transactions of the Cambridge Philosophical Society* **23**: 19–41.

Hale, B. and Wright, C. (2001), *The Reason's Proper Study: Essays towards a Neo-Fregean Philosophy of Mathematics*. New York: Oxford University Press.

Hall, C. (1988), "The 'Worthless Lands Hypothesis' and Australia's National Parks and Reserves." In K. J. Frawley and N. M. Semple (eds.), *Australia's Ever Changing Forests*. Canberra: Australian Defense Force Academy, Department of Geography and Oceanography, Special Publication Number 1.

Hallam, T. (1986), "Population Dynamics in a Homogeneous Environment." In T. Hallam and S. Levin (eds.), *Mathematical Ecology: An Introduction*. New York: Springer, pp. 61–94.

Hardin, G. (1960), "The Competitive Exclusion Principle." *Science* **131**: 1292–1297.

Harris, C. (1977), "Towards a Historical Perspective." *Proceedings of the Royal Geographical Society of Australia* **78**: 55–71.

Hastings, A. (1988), "Food Web Theory and Stability." *Ecology* **69**: 1665–1668.

Heath, J. (2015), "Methodological Individualism." In *The Stanford Encyclopedia of Philosophy* (Spring 2015 edition), E. N. Zalta (ed.), https://plato.stanford.edu/archives/spr2015/entries/methodological-individualism/.

Hempel, P. (1966), *Philosophy of Natural Science*. Englewood Cliffs, NJ: Prentice-Hall.

Hendry, R.; Needham, P.; and Woody, A. (eds.) (2011), *Handbook of the Philosophy of Science, vol. 6: Philosophy of Chemistry*. New York: Elsevier.

Herbold, B. and Moyle, P. (1986), “Introduced Species and Vacant Niches.” *The American Naturalist* **128**: 751–760.

Higgs, A. J. (1981), “Island Biogeography and Nature Reserve Design.” *Journal of Biogeography* **8**: 117–124.

Higgs, A. J. and Margules, C. (1980), “Reserve Area and Strategies for Nature Conservation.” In M. D. Hooper (ed.), *Proceedings of the Symposium on Area and Isolation*. Cambridge: Institute of Terrestrial Ecology.

Higgs, A. J. and Usher, M. B. (1980), “Should Nature Reserves Be Large or Small?” *Nature* **285**: 568–569.

Hinrichsen, D. and Pritchard, A. (2005), *Mathematical Systems Theory I*. New York: Springer.

Hirsch, M. and Smale, S. (1974), *Differential Equations, Dynamical Systems, and Linear Algebra*. New York: Academic Press.

Hoover, K. (2010), “Idealizing Reduction: The Microfoundations of Macroeconomics.” *Erkenntnis* **73**: 329–347.

Horwich, P. (1998), *Meaning*. New York: Oxford University Press.

Hubbell, S. (2001), *The Unified Neutral Theory of Biodiversity and Biogeography*. Princeton: Princeton University Press.

Hurlbert, S. (1971), “The Nonconcept of Species Diversity: A Critique and Alternative Parameters.” *Ecology* **52**: 577–586.

Hutchinson, G. E. (1944), “Limnological Studies in Connecticut. Part 7. A Critical Examination of the Supposed Relationship between Phytoplankton Periodicity and Chemical Changes in Lake Waters.” *Ecology* **25**: 3–26.

(1957), “Concluding Remarks.” *Cold Spring Harbor Symposia on Quantitative Biology* **22**: 415–427.

(1959), “Homage to Santa Rosalia, or, Why Are There So Many Kinds of Animals?” *American Naturalist* **93**: 145–159.

(1961), “The Paradox of the Plankton.” *American Naturalist* **95**: 137–145.

(1965), *The Ecological Theater and the Evolutionary Play*. New Haven, CT: Yale University Press.

(1978), *An Introduction to Population Ecology*. New Haven, CT: Yale University Press.

International Union for Conservation of Nature and Natural Resources (1980), *World Conservation Strategy: Living Resource Conservation for Sustainable Development*. Gland: IUCN.

Ismael, J. (2015), “Quantum Mechanics ” In *The Stanford Encyclopedia of Philosophy* (Spring 2015 edition), E. N. Zalta (ed.), https://plato.stanford.edu/archives/spr2015/entries/qm/.

Ive, J. R. and Cocks, K. D. (1983), "SIRO-PLAN and LUPLAN: An Australian Approach to Land Use Planning. 2. The LUPLAN Land Use Planning Package." *Environment and Planning B* **10**: 347–356.

Ive, J. R.; Davis, J. R.; and Cocks, K. D. (1985), "LUPLAN: A Computer Package to Support Inventory, Evaluation, and Allocation of Land Resources." *Soil Survey and Land Evaluation* **5**: 77–87.

Jamieson, D. (2014), *Reason in a Dark Time: Why the Struggle against Climate Change Failed – And What It Means for Our Future*. New York: Oxford University Press.

Janzen, D. (1970), "Herbivores and the Number of Tree Species in Tropical Forests." *The American Naturalist* **104**: 501–528.

(1986), "The Future of Tropical Ecology." *Annual Review of Ecology and Systematics* **17**: 305–324.

Jeffrey, R. (1956), "Valuation and Acceptance of Scientific Hypotheses." *Philosophy of Science* **33**: 237–246.

Justus, J. (2005), "Qualitative Scientific Modeling and Loop Analysis." *Philosophy of Science* **72**: 1272–1286.

(2006), "Loop Analysis and Qualitative Modeling: Limitations and Merits." *Biology and Philosophy* **21**: 647–666.

(2008a), "Complexity, Diversity, Stability." In S. Sarkar and A. Plutynski (eds.), *A Companion to the Philosophy of Biology*. Malden, MA: Blackwell, pp. 321–350.

(2008b), "Ecological and Lyapunov Stability." *Philosophy of Science* **75**: 421–436.

(2011), "A Case Study in Concept Determination: Ecological Diversity." In K. de Laplante, B. Brown, and K. Peacock (eds.), *Handbook of the Philosophy of Science, vol. 11: Philosophy of Ecology*. Amsterdam: Elsevier, pp. 147–168.

(2012a), "The Elusive Basis of Inferential Robustness." *Philosophy of Science* **79**: 795–807.

(2012b), "Carnap on Concept Determination: Methodology for Philosophy of Science." *European Journal for Philosophy of Science* **2**: 161–179.

(2013), "Philosophical Issues in Ecology." In K. Kampourakis (ed.), *Philosophy of Biology: A Companion for Educators*. New York: Springer, pp. 343–371.

(2014), "Methodological Individualism in Ecology." *Philosophy of Science* **81**: 770–784.

(2019), "Ecology and the Superfluous Niche." *Philosophical Topics* **47**: 105–123.

Justus, J. and Sarkar, S. (2002), "The Principle of Complementarity in the Design of Reserve Networks to Conserve Biodiversity: A Preliminary History." *Journal of Biosciences* **27**: 421–435.

Justus, J.; Colyvan, M.; Regan, H.; and Maguire, L. (2009a), "Buying into Conservation: Intrinsic versus Instrumental Value." *Trends in Ecology and Evolution* **24**: 187–191.

(2009b), "Response to Sagoff." *Trends in Ecology and Evolution* **24**: 644.

Kahneman, D. and Tversky, A. (eds.) (2000), *Choices, Values, and Frames*. New York: Cambridge University Press.

Kangas, P. C. (1986), "A Method for Predicting Extinction Rates Due to Deforestation in Tropical Life Zones." Abstract. In *International Congress of Ecology Meeting Program*. Ithaca, NY: Ecological Society of America, p. 194.

Kepler, C. B. and Scott, J. M. (1985), "Conservation of Island Ecosystems." In P. J. Moors (ed.), *Conservation of Island Birds*. Norwich: Paston Press, pp. 255–271.

Kerferd, G. B. (1981), *The Sophistic Movement*. New York: Cambridge University Press.

Kim, J. (1998), *Mind in a Physical World: An Essay on the Mind–Body Problem and Mental Causation*. Cambridge, MA: MIT Press.

Kimura, M. (1983), *The Neutral Theory of Molecular Evolution*. Cambridge: Cambridge University Press.

Kincaid, H. (1997), *Individualism and the Unity of Science*. New York: Rowman and Littlefield.

(2004), "Methodological Individualism and Economics." In J. Davis, A. Marciano, and J. Runde (eds.), *The Elgar Companion to Economics and Philosophy*. Cheltenham: Edward Elgar, pp. 299–314.

Kincaid, H. and Ross, D. (eds.) (2017), *The Oxford Handbook of Philosophy of Economics*. New York: Oxford University Press.

Kincaid, H.; Dupré, J.; and Wylie, A. (eds.) (2007), *Value Free Science: Ideal or Illusion?* New York: Oxford University Press.

Kindlmann, P. (2007), "Inverse Latitudinal Gradients in Species Diversity." In D. Storch and P. A. Marquet (eds.), *Scaling Biodiversity*. New York: Cambridge University Press, pp. 246–257.

Kingsland, S. (1995), *Modeling Nature: Episodes in the History of Population Ecology*. 2nd edition. Chicago: University of Chicago Press.

(2002a), "Creating a Science of Nature Reserve Design: Perspectives from History." *Environmental Modeling and Assessment* **7**: 61–69.

(2002b), "Designing Nature Reserves: Adapting Ecology to Real-World Problems." *Endeavor* **26**: 9–14.

Kirkpatrick, J. (1983), "An Iterative Method for Establishing Priorities for the Selection of Nature Reserves: An Example from Tasmania." *Biological Conservation* **25**: 127–134.

Kirkpatrick, J.; Brown, M. J.; and Moscal, A. (1980), *Threatened Plants of the Tasmanian Central East Coast*. Hobart: Tasmanian Conservation Trust.

Kleiber, M. (1932), "Body Size and Metabolism." *Hilgardia* **6**: 315–353.

Koertge, N. (ed.) (2005), *Scientific Values and Civic Virtues*. New York: Oxford University Press.

Kohn, D. (2009), "Darwin's Keystone: The Principle of Divergence." In M. Ruse and R. Richards (eds.), *Cambridge Companion to the "Origin of Species."* Cambridge: Cambridge University Press, pp. 87–108.

Kot, M. (2001), *Elements of Mathematical Ecology*. New York: Cambridge University Press.

Krebs, C. (1989), *Ecological Methodology*. New York: Harper and Row.

(2001), *Ecology*. New York: Benjamin Cummings.

Kuhn, T. (1962), *The Structure of Scientific Revolutions*. Chicago: University of Chicago Press.

(1977), *The Essential Tension. Selected Studies in Scientific Tradition and Change*. Chicago: University of Chicago Press.

Kushlan, J. A. (1979), "Design and Management of Continental Wildlife Reserves: Lessons from the Everglades." *Biological Conservation* **15**: 281–290.

Lack, D. (1954), *The Natural Regulation of Animal Numbers*. New York: Oxford University Press.

Lande, R. (1988), "Demographic Models of the Northern Spotted Owl (*Strix occidentalis caurina*)." *Oecologia* **75**: 601–607.

(1996), "Statistics and Partitioning of Species Diversity, and Similarity among Multiple Communities." *Oikos* **76**: 5–13.

Lange, M. (2005), "Ecological Laws: What Would They Be and Why Would They Matter?" *Oikos* **110**: 394–403.

Lawton, J. (1982), "Vacant Niches and Unsaturated Communities: A Comparison of Bracken Herbivores at Sites on Two Continents." *Journal of Animal Ecology* **51**: 573–595.

(1984), "Non-competitive Populations, Non-convergent Communities, and Vacant Niches: The Herbivores of Bracken." In D. Strong, D. Simberloff, L. Abele, and A. Thistle (eds.), *Ecological Communities: Conceptual Issues and the Evidence*. Princeton: Princeton University Press, pp. 67–101.

(1999), "Are There General Laws in Ecology?" *Oikos* **84**: 177–192.

Levi, I. (1960), "Must the Scientist Make Value Judgments?" *Journal of Philosophy* **58**: 345–357.

(1962), "On the Seriousness of Mistakes." *Philosophy of Science* **29**: 47–65.

Levins, R., and Lewontin, R. (1985), *The Dialectical Biologist*. Cambridge, MA: Harvard University Press.

Lewontin, R. (1969), "The Meaning of Stability." In G. Woodwell and H. Smith (eds.), *Diversity and Stability in Ecological Systems*. Brookhaven, NY: Brookhaven Laboratory, Publication No. 22, pp. 13–24.

(1970), "The Units of Selection." *Annual Review of Ecology and Systematics* **1**: 1–18.

(1972), "The Apportionment of Human Diversity." *Evolutionary Biology* **6**: 381–398.

Linquist, S (2008), "But Is It Progress? On the Alleged Advances of Conservation Biology over Ecology." *Biology and Philosophy* **23**: 529–544.

Logofet, D. (1993), *Matrices and Graphs: Stability Problems in Mathematical Ecology*. Ann Arbor, MI: CRC Press.

Łomnicki, A. (1978), "Individual Differences between Animals and the Natural Regulation of Their Numbers." *Journal of Animal Ecology* **47**: 461–475.

Longino, H. (1996), "Cognitive and Non-cognitive Values in Science: Rethinking the Dichotomy." In L. H. Nelson and J. Nelson (eds.), *Feminism, Science, and the Philosophy of Science*. New York: Kluwer, pp. 39–58.

Lookwood, D. (2008), "When Logic Fails Ecology." *Quarterly Review of Biology* **83**: 57–64.

Loreau, M. (2010), *From Populations to Ecosystems: Theoretical Foundations for a New Ecological Synthesis*. Princeton: Princeton University Press.

Lotka, A. (1932), "The Growth of Mixed Populations: Two Species Competing for a Common Food Supply." *Journal of the Washington Academy of Sciences* **22**: 461–469.

Lyapunov, A. ([1892], 1992), *The General Problem of the Stability of Motion*. London: Taylor and Francis.

MacArthur, R. (1955), "Fluctuations of Animal Populations, and a Measure of Community Stability." *Ecology* **36**: 533–536.

(1958), "Population Ecology of Some Warblers of Northeastern Coniferous Forests." *Ecology* **39**: 599–619.

(1972), *Geographical Ecology*. New York: Harper and Row.

MacArthur, R. and Wilson, E. O. (1963), "An Equilibrium Theory of Insular Zoogeography." *Evolution* **17**: 373–387.

(1967), *The Theory of Island Biogeography*. Princeton: Princeton University Press.

MacBride, F. (2003), "Speaking with Shadows: A Study of Neo-logicism." *British Journal for the Philosophy of Science* **54**: 103–163.

Machamer, P. and Wolters, G. (eds.) (2004), *Science, Values, and Objectivity*. Pittsburgh, PA: University of Pittsburgh Press.

Machery, E. (2017), *Philosophy within Its Proper Bounds*. New York: Oxford University Press.

Mackie, J. L. (1977), *Ethics: Inventing Right and Wrong*. New York: Penguin Books.

Maclaurin, J. and Sterelny, K. (2008), *What Is Biodiversity?* Chicago: University of Chicago Press.

Magurran, A. E. (1988), *Ecological Diversity and Its Measurement*. Princeton: Princeton University Press.

Margules, C. (1989a), "Introduction to Some Australian Developments in Conservation Evaluation." *Biological Conservation* **50**: 1–11.

(1989b), "Selecting Nature Reserves in South Australia." In J. Noble and R. Bradstock (eds.), *Mediterranean Landscapes in Australia: Mallee Ecosystems and Their Management*. Melbourne: CSIRO, pp. 398–406.

Margules, C. and Sarkar, S. (2007), *Systematic Conservation Planning*. New York: Cambridge University Press.

Margules, C. and Stein, J. L. (1989), "Patterns in the Distributions of Species and the Selection of Nature Reserves: An Example from Eucalyptus Forests in South-eastern New South Wales." *Biological Conservation* **50**: 219–238.

Margules, C.; Higgs, A. J.; and Rafe, R. W. (1982), "Modern Biogeographic Theory: Are There Lessons for Nature Reserve Design?" *Biological Conservation* **24**: 115–128.

Margules, C.; Nicholls, A. O.; and Pressey, R. (1988), "Selecting Networks of Reserves to Maximise Biological Diversity." *Biological Conservation* **43**: 63–76.

Margurann, A. (2004), *Measuring Biological Diversity*. Malden, MA: Blackwell Publishing.

Marquet, P.; Quiñones, R.; Abades, S.; Labra, F.; Tognelli, M.; Arim, M.; and Rivadeneira, M. (2005), "Scaling and Power-Laws in Ecological Systems." *The Journal of Experimental Biology* **208**: 1749–1769.

May, R. (1974), *Stability and Complexity in Model Ecosystems*. 2nd edition. Princeton: Princeton University Press.

Mayo, D. (1991), "Sociological versus Metascientific View of Risk Assessment." In D. Mayo and R. Hollander (eds.), *Acceptable Evidence: Science and Values in Risk Management*. New York: Oxford University Press, pp. 249–279.

McCarthy M. (2007) *Bayesian Methods for Ecology*. New York: Cambridge University Press.

McHarg, I. L. (1969), *Design with Nature*. Garden City, NY: Doubleday.

McIntosh, R. (1985), *The Background of Ecology: Concept and Theory*. New York: Cambridge University Press.

McKendry, J. E. and Machlis, G. E. (1991), "The Role of Geography in Extending Biodiversity GAP Analysis." *Applied Geography* **11**: 135–152.

McLaughlin, B. and Bennett, K. (2018), "Supervenience." In *The Stanford Encyclopedia of Philosophy* (Winter 2018 edition), E. N. Zalta (ed.), https://plato.stanford.edu/archives/win2018/entries/supervenience/.

Meffe, G. K. and Carroll, C. R. (eds.) (1994), *Principles of Conservation Biology*. Sunderland: Sinauer Associates.

Mehdiabadi, N. and Gilbert, L. (2002), Colony-Level Impacts of Parasatoid Flies on Fire Ants." *Proceedings of the Royal Society of London B* **269**: 1695–1699.

Mellor, D. H. (2005), *Probability: A Philosophical Introduction*. London: Routledge.

Mikkelson, G. (2003), "Ecological Kinds and Ecological Laws." *Philosophy of Science* **70**: 1390–1400.

Millstein, R. (2009), "Populations as Individuals." *Biological Theory* **4**: 267–273.

(2013), "Exploring the Status of Population Genetics: The Role of Ecology." *Biological Theory* **7**: 346–357.

Mitchell, S. (2009), *Unsimple Truths: Science, Complexity, and Policy*. Chicago: University of Chicago Press.

Møller, A. P. (1998), "Evidence of Larger Impact of Parasites on Hosts in the Tropics: Investment in Immune Function within and outside the Tropics?" *Oikos* **82**: 265–270.

Murphy, D. (2015), "Concepts of Disease and Health." In *The Stanford Encyclopedia of Philosophy* (Spring 2015 edition), E. N. Zalta (ed.), https://plato.stanford.edu/archives/spr2015/entries/health-disease/.

Myers, N. (1988), "Threatened Biotas: 'Hot Spots' in Tropical Forests." *The Environmentalist* **8**: 187–208.

Nagel, T. (1974), "What Is It like to Be a Bat?" *Philosophical Review* **83**: 435–450.

Nelkin, D. (1977), "Scientists and Professional Responsibility: The Experience of American Ecologists." *Social Studies of Science* **7**: 75–95.

Nicholls, A. O. and Margules, C. (1991), "The Design of Studies to Demonstrate the Biological Importance of Corridors." In D. A. Saunders and R. J. Hobbs (eds.), *Nature Conservation 2: The Role of Corridors*. Chipping Norton: Surrey Beaty and Sons, pp. 49–61.

(1993), "An Upgraded Place-Prioritization Algorithm." *Biological Conservation* **64**: 165–169.

Niiniluoto, I. (1993), "The Aim and Structure of Applied Research." *Erkenntnis* **38**: 1–21.

(2013), "Values in Design Science." *Studies in History and Philosophy of Science Part A* **46**: 11–15.

Norton, B. (1987), *Why Preserve Natural Variety?* Princeton: Princeton University Press.

(1988), "Commodity, Amenity, and Morality." In E. O. Wilson (ed.), *Biodiversity*. Washington, DC: National Academy Press.

(1994), *Towards Unity among Environmentalists*. New York: Oxford University Press.

(2005), *Sustainability: A Philosophy of Adaptive Ecosystem Management*. Chicago: University of Chicago Press.

(2006), "Towards a Policy-Relevant Definition of Biodiversity." In J. M. Scott, D. Goble, and F. W. Davis (eds.), *The Endangered Species Act at Thirty*, vol. 2. New York: Island Press.

Noss, R. F. (1986), "Dangerous Simplifications in Conservation Biology." *Bulletin of the Ecological Society of America* **67**: 278–279.

Nozick, R. (1977), "An Austrian Methodology." *Synthese* **36**: 353–392.

Oberdan, T. (1990), "Positivism and the Pragmatic Theory of Observation." *Proceedings of the Biennial Meeting of the Philosophy of Science Association* **1**: 25–37.

Odenbaugh, J. (2001), "Ecological Stability, Model Building, and Environmental Policy: A Reply to Some of the Pessimism." *Philosophy of Science* (Proceedings) **68**: S493–S505.

(2005), "Idealized, Inaccurate but Successful: A Pragmatic Approach to Evaluating Models in Theoretical Ecology." *Biology and Philosophy* **20**: 231–255.

(2007), "Seeing the Forest and the Trees." *Philosophy of Science* **74**: 628–641.

(2009), "Review of S. Sarkar, *Biodiversity and Environmental Philosophy: An Introduction*." *Biology and Philosophy* **24**: 541–550.

(2019), *Ecological Models*. New York: Cambridge University Press.

(forthcoming), "Neutrality, Niche, and Nulls: Causal Relevance in Ecology." In C. Waters and J. Woodward (eds.), *Philosophical Perspectives on Causal Reasoning in Biology*. University of Minnesota Press.

Odling-Smee, F.; Laland, K.; and Feldman, M. (2003), *Niche Construction: The Neglected Process in Evolution*. Princeton: Princeton University Press.

O'Hara, R. (2005), "The Anarchist's Guide to Ecological Theory, or, We Don't Need No Stinkin' Laws." *Oikos* **110**: 390–393.

Okasha, S. (2011), "Optimal Choice in the Face of Risk: Decision Theory Meets Evolution." *Philosophy of Science* **78**: 83–104.

Oksanen, L.; Fretwell, S.; Arrunda, J.; and Niemela, P. (1981), "Exploitation Ecosystems in Gradients of Primary Productivity." *American Naturalist* **118**: 240–261.

Oppenheimer, M.; Oreskes, N.; Jamieson, D.; Brysse, K.; O'Reilly, J.; Shindell, M.; and Wazeck, M. (2019), *Discerning Experts: The Practices of Scientific Assessment for Environmental Policy*. Chicago: University of Chicago Press.

Pearce, T. (2010), "'A Great Complication of Circumstances': Darwin and the Economy of Nature." *Journal of the History of Biology* **43**: 493–528.

Peet, R. K. (1974), "The Measurement of Species Diversity." *Annual Review of Ecology and Systematics* **5**: 285–307.

Peixoto, M. M. (1959), "On Structural Stability." *Annals of Mathematics* **69**: 199–222.

Peters, R. (1991), *A Critique for Ecology*. New York: Cambridge University Press.

Pettigrew, R. (2016), *Accuracy and the Laws of Credence*. New York: Oxford University Press.

Pianka, E. (1966), "Latitudinal Gradients in Species Diversity: A Review of Concepts." *American Naturalist* **100**: 33–46.

(2000), *Evolutionary Ecology*. New York: Addison Wesley Longman.

Pickett, S. T. A. and Thompson, J. N. (1978), "Patch Dynamics and the Design of Nature Reserves." *Biological Conservation* **13**: 27–37.

Pielou, E. C. (1977), *Mathematical Ecology*. 2nd edition. New York: John Wiley & Sons.

Pimm, S. (1979), "Complexity and Stability: Another Look at MacArthur's Original Hypothesis." *Oikos* **33**: 351–357.

(1984), "The Complexity and Stability of Ecosystems." *Ecology* **61**: 219–225.

(1991), *The Balance of Nature?* Chicago: University of Chicago Press.

Pressey, R. (1990a), "Reserve Selection in New South Wales: Where to from Here?" *Australian Zoologist* **26**: 70–75.

(1990b), "Clearing and Conservation in the Western Division." *National Parks Journal* **34**: 16–24.

(1992), "Opportunism in Acquiring Land for Reserves: Why It's a Bad Idea." *National Parks Journal* **36**: 19–22.

(1993), "What You Can Save Depends on What You Know: Why Research on Reserve Selection Is Vital." *National Parks Journal* (June): 12–14.

(1994), "Ad Hoc Reservations: Forward or Backward Steps in Developing Reprentative Reserve Systems?" *Conservation Biology* **8**: 662–668.

(1998), "Algorithms, Politics, and Timber: An Example of the Role of Science in a Public, Political Negotiation Process over New Conservation Areas in Production Forests." In R. T. Wills, R. I. Hobbs, and M. D. Fox (eds.), *Ecology for Everyone: Communicating Ecology to Scientists, the Public and the Politicians*. Chipping Norton: Surrey Beatty and Sons, pp. 73–87.

(2002), "Classics in Physical Geography Revisited." *Progress in Physical Geography* **26**: 434–441.

Pressey, R. and Nicholls, A. O. (1989), "Application of a Numerical Algorithm to the Selection of Reserves in Semi-Arid New South Wales." *Biological Conservation* **50**: 263–278.

Pressey, R. and Tulley, S. L. (1994), "The Cost of Ad Hoc Reservation: A Case Study in Western New South Wales." *Australian Journal of Ecology* **19**: 375–384.

Pressey, R.; Bedward, M.; and Nicholls, A. O. 1990. "Reserve Selection in the Mallee Lands." In J. C. Noble, P. J. Joss, and G. K. Jones (eds.), *The Mallee Lands: A Conservation Perspective*. Melbourne: CSIRO, pp. 167–178.

Pressey, R.; Johnson, I. R.; and Wilson, P. D. (1994), "Shades of Irreplaceability: Toward a Measure of the Contribution of Sites to a Reservation Goal." *Biodiversity and Conservation* **3**: 242–262.

Preston, F. W. (1962), "The Canonical Distribution of Commoness and Rarity." *Ecology* **43**: 185–215 and 410–432.

Primack, R. B. (1993), *Essentials of Conservation Biology*. Sunderland: Sinauer Associates.

Proctor, R. N. (1991), *Value-Free Science? Purity and Power in Modern Knowledge*. Cambridge, MA: Harvard University Press.

Puccia, C. and Levins, R. (1985), *Qualitative Modeling of Complex Systems: An Introduction to Loop Analysis and Time Averaging*. Cambridge, MA: Harvard University Press.

Putnam, H. (1990), *Realism with a Human Face*. Cambridge, MA: Harvard University Press.

(2002), *The Collapse of the Fact/Value Dichotomy and Other Essays*. Cambridge, MA: Harvard University Press.

RACAC (Resource and Conservation Assessment Council) (1996), *Draft Interim Forestry Assessment Report*. Sydney: RACAC.

Real, L. and Levin, S. (1991), "The Role of Theory in the Rise of Modern Ecology." In L. Real and J. Brown (eds.), *Foundations of Ecology*. Chicago: University of Chicago Press, pp. 177–191.

Rebelo, A. G. and Siegfried, W. R. (1990), "Protection of Fynbos Vegetation: Ideal and Real-World Options." *Biological Conservation* **54**: 15–31.

Recher, H. (1976), "An Ecologist's View: The Failure of Our National Parks System." *Australian Natural History* **18**: 398–405.

Regan, H.; Colyvan, M.; and Burgman, M. (2002), "A Taxonomy and Treatment of Uncertainty for Ecology and Conservation Biology." *Ecological Applications* **12**: 618–628.

Reydon, T. (2013), "Classifying Life, Reconstructing History and Teaching Diversity: Philosophical Issues in the Teaching of Biological Systematics and Biodiversity." *Science and Education* **22**: 189–220.

Righter, R. W. (2005), *The Battle over Hetch Hetchy: America's Most Controversial Dam and the Birth of Modern Environmentalism*. New York: Oxford University Press.

Robert, C. and Casella, G. (2004), *Monte Carlo Statistical Methods*. New York: Springer.

Roberts, J. T. (2007), "Is Logical Empiricism Committed to the Ideal of Value-Free Science?" In H. Kincaid, J. Dupré, and A. Wylie (eds.), *Value-Free Science? Ideals and Illusions*. New York: Oxford University Press, pp. 143–163.

Rosenberg, A. and Bouchard, F. (2015), "Fitness." In *The Stanford Encyclopedia of Philosophy* (Fall 2015 edition), E. N. Zalta (ed.), https://plato.stanford.edu/archives/fall2015/entries/fitness/.

Rosenzweig, M. (1992), "Species Diversity Gradients: We Know More and Less than We Thought." *Journal of Mammology* **73**: 715–730.

Roughgarden, J. (2012), "Individual-Based Models in Ecology: An Evaluation." PhilSci Archive: http://philsci-archive.pitt.edu/9434/1/RoughgardenPSA2012IBMLecture .pdf.

Rudner, R. (1953), "The Scientist qua Scientist Makes Value Judgments." *Philosophy of Science* **20**: 1–6.

Runte, A. (1972), "Yellowstone: It's Useless, So Why Not a Park?" *National Parks and Conservation Magazine* (March): 4–7.

(1977), "The National Park Idea: Origins and Paradox of the American Experience." *Journal of Forest History* **21**: 61–75.

(1979), *National Parks: The American Experience*. Lincoln: University of Nebraska Press.

(1983), "Reply to Sellars." *Journal of Forest History* **27**: 135–141.

Ruse, M. (1970), "Are There Laws in Biology?" *Australasian Journal of Philosophy* **48**: 234–246.

(2013), *The Gaia Hypothesis: Science on a Pagan Planet*. Chicago: University of Chicago Press.

Sagoff, M. (2009), "Intrinsic Value: A Response to Justus et al." *Trends in Ecology and Evolution* **24**: 643.

Santana, C. (2014), "Save the Planet: Eliminate Biodiversity." *Biology and Philosophy* **29**: 761–780.

(2016), "Biodiversity Eliminativism." In J. Garson, A. Plutynski, and S. Sarkar (eds.), *Routledge Handbook for the Philosophy of Biodiversity*. New York: Routledge, pp. 86–95.

(2018), "Biodiversity Is a Chimera and Chimeras Aren't Real." *Biology and Philosophy* 33:15, https://doi.org/10.1007/s10539-018-9626-2.

Sarkar, S. (1996), "Ecological Theory and Anuran Declines." *Bioscience* **46**: 199–207.

(2002), "Defining 'Biodiversity'; Assessing Biodiversity." *Monist* **85**: 131–155.

(2005), *Biodiversity and Environmental Philosophy: An Introduction*. Cambridge: Cambridge University Press.

(2007), "From Ecological Diversity to Biodiversity." In D. Hull and M. Ruse (eds.), *Cambridge Companion to the Philosophy of Biology*. Cambridge: Cambridge University Press.

Sarkar, S.; Justus, J; Fuller, T.; Kelley; C.; Garson, J.; and Mayfield, M. (2005), "Effectiveness of Estimator Surrogates for the Selection of Conservation Area Networks." *Conservation Biology* **19**: 815–825.

Sarkar, S.; Pressey, R. L.; Faith, D. P.; Margules, C.; Fuller, T.; Stoms, D.; Moffett, A.; Wilson, K. A.; Williams, K. J.; Williams, P. H.; and Andelman, S. (2006),

"Biodiversity Conservation Planning Tools: Present Status and Challenges for the Future." *Annual Review of Environment and Resources* **31**: 123–159.

Schaffner, K. (1969a), "The Watson–Crick Model and Reductionism." *British Journal for the Philosophy of Science* **20**: 325–348.

(1969b), "Chemical Systems and Chemical Evolution: The Philosophy of Molecular Biology." *American Scientist* **57**: 410–420.

Schoener, T. (1989), "The Ecological Niche." In J. Cherrett (ed.), *Ecological Concepts: The Contribution of Ecology to an Understanding of the Natural World*. Oxford: Blackwell, pp. 79–113.

Scott, J. M.; Csuti, B.; Jacobi, J. D.; and Estes, J. E. (1987), "Species Richness: A Geographical Approach to Protecting Future Biological Diversity." *Bioscience* **37**: 782–788.

Scott, J. M.; Csuti, B.; and Caicco, S. (1991a), "GAP Analysis: Assessing Protection Needs." In W. E. Hudson (ed.), *Landscape Linkages and Biodiversity*. Washington, DC: Island Press, pp. 15–26.

Scott, J. M.; Csuti, B.; Smith, K.; Estes, J. E.; and Caicco, S. (1991b), "GAP Analysis of Species Richness and Vegetation Cover: An Integrated Biodiversity and Conservation Strategy." In K. A. Kohm (ed.), *Balancing on the Brink of Extinction*. Washington, DC: Island Press, pp. 282–297.

Scott, J.; Davis, F.; Csuti, B.; Noss, R.; Butterfield, B.; Groves, C.; Anderson, H.; Caicco, S.; D'erchia, F.; Edwards, T. C., Jr.; Ulliman, J.; and Wright, R. G. (1993), "GAP Analysis: A Geographic Approach to Protection of Biological Diversity." *Wildlife Monographs* **123**: 1–41.

Scudo, Francesco (1971), "Vito Volterra and Theoretical Ecology." *Theoretical Population Biology* **2**: 1–23.

Sen, A. (2009), *The Idea of Justice*. Cambridge, MA: Harvard University Press.

Shrader-Frechette, K. (1990), "Island Biogeography, Species–Area Curves, and Statistical Errors: Applied Biology and Scientific Rationality." *Proceedings of the Philosophy of Science Association* **1**: S447–S456.

Shrader-Frechette, K. S. (1991), *Risk and Rationality: Philosophical Foundations for Populist Reforms*. Berkeley: University of California Press.

Shrader-Frechette, K. S. and McCoy, E. D. (1993), *Method in Ecology*. New York: Cambridge University Press.

Shulz, Armin (2020), *Structure, Evidence, and Heuristic: Evolutionary Biology, Economics, and the Philosophy of their Relationship*. New York: Routledge.

Siegfried, W. R. (1978), "Let the Strandwolf Fly." *African Wildlife* **32**: 10–14.

(1989), "Preservation of Species in South African Nature Reserves." In B. J. Huntley (ed.), *Biotic Diversity in Southern Africa: Concepts and Conservation*. Cape Town: Oxford University Press, pp. 186–201.

Simberloff, D. (1986), "Design of Nature Reserves." In M. B. Usher (ed.), *Wildlife Conservation Evaluation*. London: Chapman and Hall, pp. 315–337.

(1987), "Simplification, Danger, and Ethics in Conservation Biology." *Bulletin of the Ecological Society of America* **68**: 156–157.

(2004), "Community Ecology: Is It Time to Move On?" *The American Naturalist* **163**: 787–799.

Simberloff, D. and Abele, L. (1976), "Island Biogeography and Conservation Practice." *Science* **193**: 285–286.

(1982), "Refuge Design and Island Biogeographic Theory: Effects of Fragmentation." *American Naturalist* **120**: 41–50.

(1984), "Conservation and Obfuscation: Subdivision of Reserves." *Oikos* **42**: 399–401.

Simberloff, D.; Cox, J.; and Mehlman, D. W. (1992), "Movement Corridors: Conservation Bargains or Poor Investments?" *Conservation Biology* **6**: 493–504.

Skyrms, B. (1996), *Evolution of the Social Contract*. Cambridge: Cambridge University Press.

(2010), *Signals: Evolution, Learning, and Information*. New York: Oxford University Press.

Slobodkin, L. (1980), *Growth and Regulation of Animal Populations*. 2nd edition. Mineola, NY: Dover.

Smart, J. J. C. (1959), "Can Biology Be an Exact Science?" *Synthese* 11: 359–368.

(1963), *Philosophy and Scientific Realism*. London: Routledge.

Smith, T. R.; Menon, S.; Starr, J. L.; and Estes, J. E. (1987), "Requirements and Principles for the Implementation and Construction of Large-Scale Geographic Information Systems." *International Journal of Geographical Information Systems* **1**: 13–31.

Smith, A.; Hines, H. B.; and Webber, P. (1989), *Mammals, Reptiles and Amphibians of the Rainforests of the Mount Warning Caldera Region*. Unpublished Report to the New South Wales National Parks and Wildlife Service.

Sober, E. (1986), "Problems for Environmentalism." In B. Norton (ed.), *The Preservation of Species: The Value of Biological Diversity*. Princeton: Princeton University Press, pp. 173–194.

(1987), "Does 'Fitness' Fit the Facts?" *Journal of Philosophy* **84**: 220–223.

(1997), "Two Outbreaks of Lawlessness in Recent Philosophy of Biology." *Philosophy of Science* 64, Supplement. Proceedings of the 1996 Biennial Meetings of the Philosophy of Science Association. Part II: Symposia Papers, pp. S458–S467.

(1999), "The Multiple Realizability Argument against Reductionism." *Philosophy of Science* **66**: 542–564.

(2007), "Evidence and Value Freedom." In H. Kinkaid, J. Dupré, and A. Wylie (eds.), *Value-Free Science: Ideal or Illusion?* New York: Oxford University Press, pp. 109–119.

(2008), *Evidence and Evolution: The Logic behind the Science*. New York: Cambridge University Press.

(2011), "A Priori Causal Models of Natural Selection." *Australasian Journal of Philosophy* **89**: 571–589.

Soulé, M. (1985), "What Is Conservation Biology?" *BioScience* **35**: 727–734.

Soulé, M. and Simberloff, D. (1986), "What Do Genetics and Ecology Tell Us about the Design of Nature Reserves?" *Biological Conservation* **35**: 19–40.

Stauffer, R. (1957), "Haeckel, Darwin, and Ecology." *Quarterly Review of Biology* **32**: 138–144.

Sterelny, K. (2001), "The Reality of Ecological Assemblages: A Palaeo-ecological Puzzle." *Biology and Philosophy* **16**: 437–461.

(2006), "Local Ecological Communities." *Philosophy of Science* **73**: 215–231.

Sterelny, K. and Griffiths, P. (1999), *Sex and Death*. Chicago: University of Chicago Press, pp. 253–280.

Strom, A. (1979a), "Impressions of a Developing Conservation Ethic, 1870–1930." *Parks and Wildlife* **2**: 45–53.

(1979b), "Some Events in Nature Conservation over the Last Forty Years." *Parks and Wildlife* **2**: 65–73.

Strong, D. (1980), "Null Hypotheses in Ecology." *Synthese* **43**: 271–285.

Tabery, J. (2014), *Beyond Versus: The Struggle to Understand the Interaction of Nature and Nurture*. Cambridge, MA: MIT Press.

Takacs, D. (1996), *The Idea of Biodiversity*. Baltimore, MD: Johns Hopkins University Press.

Taylor, P. (1986), *Respect for Nature: A Theory for Environmental Ethics*. Princeton: Princeton University Press.

Taylor, P. and Blum, A. (1991), "Ecosystems as Circuits: Diagrams and the Limits of Physical Analogies." *Biology and Philosophy* **6**: 275–294.

Terborgh, J. (1974), "Preservation of Natural Diversity: The Problem of Extinction Prone Species." *Bioscience* **24**: 715–722.

(1976), "Island Biogeography and Conservation: Strategy and Limitations." *Science* **193**: 1029–1030.

Thompson, P. (1995), "Managing Complexity and Dynamics: Is There a Difference between Physics and Biology." *Canadian Journal of Philosophy* **20** (supplementary volume): 275–302.

(2011), *Agro-Technology: A Philosophical Introduction*. Cambridge: Cambridge University Press.

Tilman, D. (1982), *Resource Competition and Community Structure*. Princeton: Princeton University Press.

(1999), "The Ecological Consequences of Biodiversity: A Search for General Principles." *Ecology* **80**: 1455–1474.

Tilman, D.; Kilham, S.; and Kilham, P. (1982), "Phytoplankton Community Ecology: The Role of Limiting Nutrients." *Annual Review of Ecology and Systematics* **13**: 349–372.

Tilman, D.; Reich, P.; and Knops, J. (2006), "Biodiversity and Ecosystem Stability in a Decade-Long Grassland Experiment." *Nature* **441**: 629–632.

Tobin, R. (1990), *The Expendable Future: U.S. Politics and the Protection of Biological Diversity*. Durham, NC: Duke University Press.

Tomlinson, R. (1988), "The Impact of the Transition from Analogue to Digital Cartographic Representation." *The American Cartographer* **15**: 249–261.

Turner, D. (2011), *Paleontology: A Philosophical Introduction*. New York: Cambridge University Press.

Turner, D. and Havstad, J. (2019), "Philosophy of Macroevolution." In *The Stanford Encyclopedia of Philosophy* (Summer 2019 edition), E. N. Zalta (ed.), https://plato.stanford.edu/archives/sum2019/entries/macroevolution/.

van Fraasen, B. (2008), *Scientific Representation: Paradoxes of Perspective*. New York: Oxford University Press.

van Riel, R. and Van Gulick, R. (2019), "Scientific Reduction." In *The Stanford Encyclopedia of Philosophy* (Spring 2019 edition), E. N. Zalta (ed.), https://plato.stanford.edu/archives/spr2019/entries/scientific-reduction/.

Wade, P. (2000), "Bayesian Methods in Conservation Biology." *Conservation Biology* **14**: 1308–1316.

Wakefield, J. (1992), "The Concept of Mental Disorder: On the Boundary between Biological Facts and Social Values." *American Psychologist* **47**: 373–388.

Weisberg, M. (2007), "Three Kinds of Idealization." *Journal of Philosophy* **104**: 639–659.

(2013), *Simulation and Similarity: Using Models to Understand the World*. New York: Oxford University Press.

Weisberg, M. and Reisman, K. (2008), "The Robust Volterra Principle." *Philosophy of Science* **75**: 106–131.

West, G.; Brown, J.; and Enquist, B. (1997), "A General Model for the Origin of Allometric Scaling Laws in Biology." *Science* **276**: 122–126.

White, L. (1967), "The Historical Roots of Our Ecologic Crisis." *Science* **155**: 1203–1207.

Whittaker, R. (1956), "Vegetation of the Great Smoky Mountains." *Ecological Monographs* **26**: 1–80.

Williams, B. (1986), *Ethics and the Limits of Philosophy*. Cambridge, MA: Harvard University Press.

Willig, M. R.; Kaufmann, D. M.; and Stevens, R. D. (2003), "Latitudinal Gradients of Biodiversity: Pattern, Process, Scale and Synthesis." *Annual Review of Ecology and Systematics* **34**: 273–309.

Willis, E. O. (1974), "Populations and Local Extinctions of Birds on Barro Colorado Island, Panama." *Ecological Monographs* **44**: 153–169.

(1984), "Conservation, Subdivistion of Reserves, and the Anti-dismemberment Hypothesis." *Oikos* **42**: 396–398.

(ed.) (1988), *Biodiversity*. Washington, DC: National Academy Press.

(1992), Interview. In J. Peppercorn, "Islands of Sanctuary: A History of the Theory of Island Biogeography and Its Application to Nature Reserve Design." M.A. thesis, Harvard University.

(1994), *The Naturalist*. Cambridge, MA: Harvard University Press.

Wilson, E. O. and Willis, E. O. (1975), "Applied Biogeography." In M. I. Cody and J. Diamond (eds.), *Ecology and Evolution of Communities*. Cambridge, MA: Harvard University Press, pp. 522–534.

Winsberg, E. (2018), *Philosophy and Climate Change*. New York: Cambridge University Press.

Worster, D. (1994), *Nature's Economy: A History of Ecological Ideas*. New York: Cambridge University Press.

# Index

www.ingramcontent.com/pod-product-compliance
Ingram Content Group UK Ltd.
Pitfield, Milton Keynes, MK11 3LW, UK
UKHW050112180125
453697UK00008B/147